SpringerBriefs in Electrical and Computer Engineering

Signal Processing

More information about this series at http://www.springer.com/series/11560

Matteo Testa · Diego Valsesia
Tiziano Bianchi · Enrico Magli

Compressed Sensing for Privacy-Preserving Data Processing

Matteo Testa
Department of Electronics
and Telecommunications
Politecnico di Torino
Turin, Italy

Tiziano Bianchi
Department of Electronics
and Telecommunications
Politecnico di Torino
Turin, Italy

Diego Valsesia
Department of Electronics
and Telecommunications
Politecnico di Torino
Turin, Italy

Enrico Magli
Department of Electronics
and Telecommunications
Politecnico di Torino
Turin, Italy

ISSN 2191-8112 ISSN 2191-8120 (electronic)
SpringerBriefs in Electrical and Computer Engineering
ISSN 2196-4076 ISSN 2196-4084 (electronic)
SpringerBriefs in Signal Processing
ISBN 978-981-13-2278-5 ISBN 978-981-13-2279-2 (eBook)
https://doi.org/10.1007/978-981-13-2279-2

Library of Congress Control Number: 2018954021

This Springer imprint is published by the registered company Springer Nature Singapore Pte Ltd.
The registered company address is: 152 Beach Road, #21-01/04 Gateway East, Singapore 189721, Singapore

Preface

Compressed sensing is an established technique for simultaneous signal acquisition and compression, as well as dimensionality reduction, based on representing a signal with a small number of random projections. Some applications are concerned with the recovery of the original signal from the random projections, which is possible using nonlinear optimization techniques, while other applications are mostly concerned with embedding the geometry of a set of signals, thus working directly on the random projections.

The former class of applications includes the emerging field of Internet of Things, where a multitude of devices acquires and transmits sensitive information. However, such devices are typically power-constrained or have very simple computational capabilities, while at the same time, being required to ensure the confidentiality of the transmitted messages. Consequently, conventional acquisition and information protection methods are a poor fit for this scenario. Replacing such techniques with compressed sensing is an appealing solution, as it could offer a security layer on top of signal acquisition and compression without additional cost. However, its use as a cryptosystem requires a careful theoretical analysis to ensure the soundness of the method. The latter class of applications includes problems in information retrieval and signal detection, where one is interested in the results to their query without revealing the sensitive information contained in the signal itself. Embeddings obtained with random projections are appealing as they can provide a privacy-preserving functionality without compromising the performance of the embedding in terms of accuracy or computational efficiency.

This book is a unique contribution which analyzes this exciting and timely topic by collecting many different contributions, scattered in the literature, as well as recent advances in the field, in a single resource with a unified notation. The objective of this book is to provide the reader with a comprehensive survey of the topic in an accessible format. The reader is guided in exploring the topic by first establishing a shared knowledge about compressed sensing and how it is used nowadays. Then, clear models and definitions for its use as a cryptosystem and a privacy-preserving embedding are laid down, before tackling state-of-the-art results for both applications. The reader will conclude the book having learned that the current results in terms of security of compressed techniques allow it to be a very promising solution to many practical problems of interest. We believe that this book can find a broad audience among researchers, scientists, or engineers with very diverse backgrounds, having interests in security, cryptography, and privacy in information retrieval systems. Accompanying software is made available on the authors' Web site (www.ipl.polito.it) to reproduce the experiments and techniques presented in the book. The only background required to the reader is a good knowledge of linear algebra, probability, and information theory.

Turin, Italy
July 2018

Matteo Testa
Diego Valsesia
Tiziano Bianchi
Enrico Magli

Contents

Chapter 1
Introduction

Abstract Information processing systems have been revolutionized by recent advances in several technological areas, like device miniaturization, wireless transmission, network infrastructure. Traditional information sources have been replaced by a multitude of devices with sensing capabilities, the so-called Internet of Things (IoT), including smart home devices, cars with autonomous driving functions, portable medical devices. Meanwhile, single processing and storage units have been replaced by cloud services, leading to interconnected systems that share and process huge amount of data. While this provides endless opportunities to tackle different societal needs, it poses several problems regarding the security and privacy of the involved data. This chapter introduces the challenges and the techniques discussed in the literature to address them and serves as an overview of the book.

Information processing systems have been revolutionized by recent advances in several technological areas, like device miniaturization, wireless transmission, network infrastructure. Traditional information sources have been replaced by a multitude of devices with sensing capabilities, the so-called Internet of Things (IoT), including smart home devices, cars with autonomous driving functions, portable medical devices. Meanwhile, single processing and storage units have been replaced by cloud services, leading to interconnected systems that share and process huge amount of data.

While this provides endless opportunities to tackle different societal needs, and results in obvious benefits for everyone, at the same time, it poses several problems regarding the security and privacy of the involved data. Entrusting sensitive data to remote services is prone to privacy issues, since the user usually has little or no control on the actual servers managing the computations. Many sensing devices may be equipped with very basic encryption and authentication capabilities, or no capabilities at all, being vulnerable to cybersecurity attacks.

Protecting information processing system from different security threats has been an active area of research in recent years. For example, secure signal processing techniques have been proposed for privacy-preserving cloud processing [28]. Nevertheless, such techniques are often based on costly cryptographic primitives and complex protocols, which limit their usefulness to scenarios involving critical or

M. Testa et al., *Compressed Sensing for Privacy-Preserving Data Processing*,
SpringerBriefs in Signal Processing, https://doi.org/10.1007/978-981-13-2279-2_1

highly valuable information. Moreover, it appears more and more evident that the large mass of IoT devices may not have sufficient capabilities for deploying conventional cryptographic solutions. Many of these devices are battery operated and are often left unattended, with very limited maintenance, limiting both their power consumption and computational capabilities [27].

Among the solutions able to meet the stringent requirements of IoT devices, compressed sensing (CS) can be considered as a very promising option. CS is a mature technology enabling simultaneous signal acquisition and compression, based on representing a signal with a small number of highly incoherent linear projections. The possibility of implementing CS through hardware acquisition (see, e.g., [8]) reduces the number of required sensing elements, limiting the overall power consumption [9, 10, 20]. This latter aspect makes the CS framework an excellent candidate for low-energy devices [16].

At the same time, it has been recognized that the inherent randomness in the CS acquisition process provides some secrecy guarantees. For example, in [21] the authors show that CS is computationally secure as long as the sensing matrix is used only once. Additional security properties of CS were later studied in [3–7, 18], showing that in the best possible scenario CS measurements leak only the energy of the sensed signal, and a framework for securing IoT devices through CS has been recently proposed in [17].

Providing a lightweight encryption layer for low-power devices is not the only security feature of CS. If one is not concerned with signal recovery, CS measurements can be modeled as signal embeddings projecting a signal in a low dimensional space in which distances are approximately preserved [12]. Due to the properties of embeddings, some authors suggest that CS can implicitly provide a privacy preserving layer enabling simple processing tasks [1], like privacy preserving data mining [15], sparse regression [34], or achieving differential privacy [14]. Several applications relying on privacy properties of CS measurements have been recently proposed, including outsourcing image data to the cloud for privacy-preserving data mining and image retrieval [29, 31, 32], generating a robust image hash [23], providing biometric template protection [2, 19, 24, 25], and implementing physical unclonable functions [11, 22, 30]. The ability to perform basic signal processing operations on confidential data is also beneficial for IoT devices, for example to detect anomalies [26].

In this book, we will try to present the vast amount of literature on the security of CS under a unifying framework. The scenario we are referring to is that exemplified in Fig. 1.1. We assume that a number of low-power sensor nodes are transmitting privacy-sensitive data to a cloud service for enabling several information processing tasks. Different users can interact with the cloud to obtain the results of the different tasks. In the above scenario, we identify two weaknesses that can be targeted by adversaries. First, the communication channel between the sensor node and the cloud can be attacked by an eavesdropper trying to get access to sensitive information. Second, the cloud can include some non-trusted entities that observe the collected information and use this knowledge for malicious purposes.

In the first case, security is achieved using the CS framework as a lightweight cryptosystems providing some level of secrecy. While here we focus on the

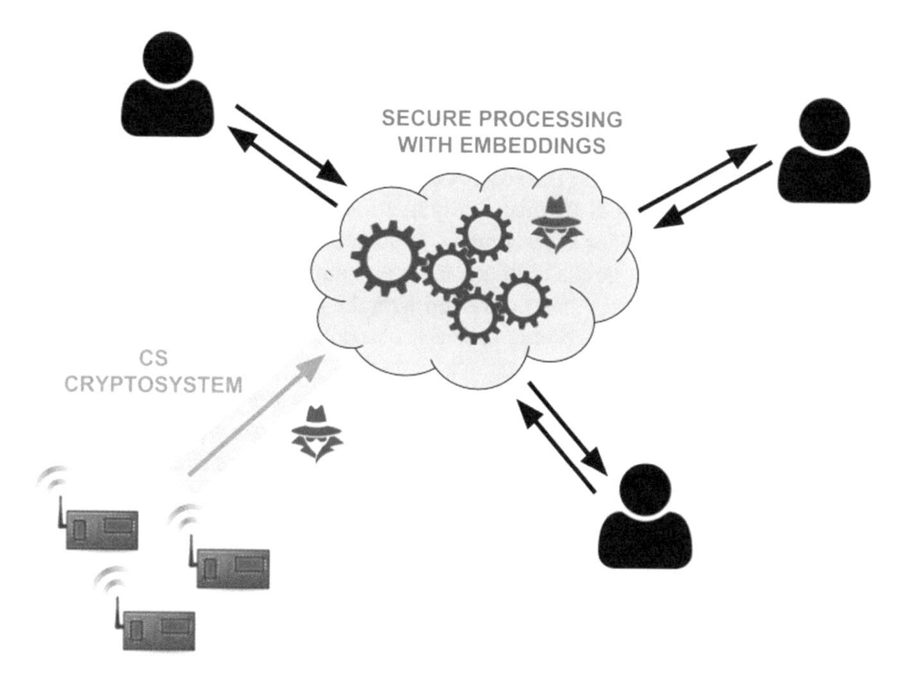

Fig. 1.1 Compressed sensing for privacy preserving data processing

confidentiality of data, it is worth noting that the envisaged CS cryptosystem acts as a private key cryptosystem, that can be easily converted in an authentication mechanism by using message authentication codes [13, 33]. In the second case, secure processing is enabled by the privacy preserving properties of CS measurements, that can be considered as secure embeddings of the underlying signals.

The materials we are going to present are organized in the following chapters. In Chap. 2, we briefly review the CS framework, discussing the acquisition model, the conditions under which the signal can be recovered, and the main reconstruction algorithms. Then, we show how CS is essentially analogous to a private key cryptosystem if signal acquisition, signal recovery, and sensing matrix generation are interpreted as encryption, decryption, and key generation functions. The basic security properties of this CS cryptosystem under different attack scenarios are discussed according to standard security definitions, identifying the scenarios that will be analyzed more in depth in the following chapter. In the second part of Chap. 2, we introduce the concept of signal embeddings, which can be seen as a generalization of CS measurements. The properties of some of the most common embeddings are briefly reviewed, then we discuss how embeddings can provide privacy-preserving functionalities in particular settings.

Chapter 3 presents the most relevant results on CS used as a cryptosystem. First, we analyze the statistical properties of CS measurements, showing that they always convey at least the energy of the sensed signal. Then, we discuss the secrecy achievable by different sensing matrix constructions. For sensing matrices made of Gaussian i.i.d. entries, we have the highest secrecy guarantees, where only the energy of the signal can be revealed. This particular case is analyzed by introducing a secrecy metric that depends on the ability to estimate the signal energy by an adversary who observes only the signal measurements. The secrecy achievable by generic sensing matrices is analyzed by introducing a distinguishability metric inspired by the standard statistical secrecy definition used in cryptography. Results are provided for matrices made of i.i.d entries with generic distributions and circulant matrices. At the end of the chapter, we discuss several issues connected with the practical implementation of a CS cryptosystem, including sensing matrix generation and quantization of sensing matrix entries.

In Chap. 4, we illustrate main results on privacy-preserving embeddings. Here, security properties of embeddings are analyzed by considering two possible scenarios for their use. In the first case, a client submits a query containing sensitive information to a server, which should respond to the query without gaining access to the private information. This is discussed describing an authentication system in which a client submit an embedding of a physical characteristic of a device, and a verification server is able to match the embedding without revealing the actual physical characteristic. Interestingly, in this case the security properties of the embedding permit to combine it with existing biometric template mechanisms, enhancing the security of the system. In the second case, a large amount of sensitive data is stored in the cloud and a user should be able to make specific queries to the cloud without gaining access to the data. Here, we describe a universal embedding that preserves distances only locally. If data are stored in the cloud using this embedding, a user is able to retrieve data close to the query, but the complete geometry of the dataset remains hidden by the embedding and data cannot be recovered.

Finally, Chap. 5 summarizes the main results discussed in the book, providing some discussion on open issues and promising avenues for future research on this topic.

References

1. Abdulghani, A., Rodriguez-Villegas, E.: Compressive sensing: from compressing while sampling to compressing and securing while sampling. In: 2010 Annual International Conference of the IEEE Engineering in Medicine and Biology Society (EMBC), pp. 1127–1130 (2010)
2. Anzaku, E.T., Sohn, H., Ro, Y.M.: Multi-factor authentication using fingerprints and user-specific random projection. In: 2010 12th International Asia-Pacific Web Conference (APWEB), pp. 415–418 (2010)
3. Bianchi, T., Bioglio, V., Magli, E.: On the security of random linear measurements. In: 2014 IEEE International Conference on Acoustics, Speech and Signal Processing (ICASSP'14), pp. 3992–3996 (2014)
4. Bianchi, T., Bioglio, V., Magli, E.: Analysis of one-time random projections for privacy preserving compressed sensing. IEEE Trans. Inf. Forensics Secur. **11**(2), 313–327 (2016)

5. Cambareri, V., Haboba, J., Pareschi, F., Rovatti, H., Setti, G., Wong, K.W.: A two-class information concealing system based on compressed sensing. In: 2013 IEEE International Symposium on Circuits and Systems (ISCAS), pp. 1356–1359 (2013)
6. Cambareri, V., Mangia, M., Pareschi, F., Rovatti, R., Setti, G.: Low-complexity multiclass encryption by compressed sensing. IEEE Trans. Signal Process. **63**(9), 2183–2195 (2015)
7. Cambareri, V., Mangia, M., Pareschi, F., Rovatti, R., Setti, G.: On known-plaintext attacks to a compressed sensing-based encryption: a quantitative analysis. IEEE Trans. Inf. Forensics Secur. **10**(10), 2182–2195 (2015)
8. Duarte, M.F., Davenport, M.A., Takhar, D., Laska, J.N., Sun, T., Kelly, K.F., Baraniuk, R.G.: Single-pixel imaging via compressive sampling. IEEE Signal Process. Mag. **25**(2), 83–91 (2008)
9. Gangopadhyay, D., Allstot, E.G., Dixon, A.M., Natarajan, K., Gupta, S., Allstot, D.J.: Compressed sensing analog front-end for bio-sensor applications. IEEE J. Solid-State Circuits **49**(2), 426–438 (2014)
10. Herman, M.A., Strohmer, T.: High-resolution radar via compressed sensing. IEEE Trans. Signal Process. **57**(6), 2275–2284 (2009)
11. Holotyak, T., Voloshynovskiy, S., Koval, O., Beekhof, F.: Fast physical object identification based on unclonable features and soft fingerprinting. In: 2011 IEEE International Conference on Acoustics, Speech and Signal Processing (ICASSP), pp. 1713–1716 (2011)
12. Johnson, W.B., Lindenstrauss, J.: Extensions of Lipschitz mappings into a Hilbert space. Contemp. Math. **26** (1984)
13. Katz, J., Lindell, Y.: Introduction to Modern Cryptography. Chapman & Hall/CRC Cryptography and Network Security Series. Chapman & Hall/CRC, London (2007)
14. Li, Y.D., Zhang, Z., Winslett, M., Yang, Y.: Compressive mechanism: utilizing sparse representation in differential privacy. In: Proceedings of the 10th Annual ACM Workshop on Privacy in the Electronic Society, WPES'11, pp. 177–182. ACM, New York (2011)
15. Liu, K., Kargupta, H., Ryan, J.: Random projection-based multiplicative data perturbation for privacy preserving distributed data mining. IEEE Trans. Knowl. Data Eng. **18**(1), 92–106 (2006)
16. Mangia, M., Marchioni, A., Pareschi, F., Rovatti, R., Setti, G.: Administering quality-energy trade-off in IOT sensing applications by means of adapted compressed sensing. IEEE J. Emerg. Sel. Top. Circuits Syst. 1–1 (2018)
17. Mangia, M., Pareschi, F., Rovatti, R., Setti, G.: Low-cost security of IOT sensor nodes with rakeness-based compressed sensing: statistical and known-plaintext attacks. IEEE Trans. Inf. Forensics Secur. **13**(2), 327–340 (2018)
18. Orsdemir, A., Altun, H., Sharma, G., Bocko, M.: On the security and robustness of encryption via compressed sensing. In: 2008 IEEE Military Communications Conference (MILCOM 2008), pp. 1–7 (2008)
19. Pillai, J.K., Patel, V.M., Chellappa, R., Ratha, N.K.: Secure and robust IRIS recognition using random projections and sparse representations. IEEE Trans. Pattern Anal. Mach. Intell. **33**(9), 1877–1893 (2011)
20. Quinsac, C., Basarab, A., Girault, J.M., Kouamé, D.: Compressed sensing of ultrasound images: sampling of spatial and frequency domains. In: 2010 IEEE Workshop on Signal Processing Systems (SIPS), pp. 231–236. IEEE (2010)
21. Rachlin, Y., Baron, D.: The secrecy of compressed sensing measurements. In: 2008 46th Annual Allerton Conference on Communication, Control, and Computing, pp. 813–817. IEEE (2008)
22. Shariati, S., Jacques, L., Standaert, F.X., Macq, B., Salhi, M.A., Antoine, P.: Randomly driven fuzzy key extraction of unclonable images. In: 2010 IEEE International Conference on Image Processing, pp. 4329–4332 (2010)
23. Sun, R., Zeng, W.: Secure and robust image hashing via compressive sensing. Multimed. Tools Appl. **70**(3), 1651–1665 (2014)
24. Teoh, A.B.J., Goh, A., Ngo, D.C.L.: Random multispace quantization as an analytic mechanism for biohashing of biometric and random identity inputs. IEEE Trans. Pattern Anal. Mach. Intell. **28**(12), 1892–1901 (2006)

25. Teoh, A.B.J., Yuang, C.T.: Cancelable biometrics realization with multispace random projections. IEEE Trans. Syst. Man Cybern. Part B (Cybern.) **37**(5), 1096–1106 (2007)
26. Testa, M., Bianchi, T., Magli, E.: Energy obfuscation for compressive encryption and processing. In: 2017 IEEE Workshop on Information Forensics and Security (WIFS), pp. 1–6 (2017)
27. Trappe, W., Howard, R., Moore, R.S.: Low-energy security: limits and opportunities in the internet of things. IEEE Secur. Priv. **13**(1), 14–21 (2015)
28. Troncoso-pastoriza, J.R., Perez-Gonzalez, F.: Secure signal processing in the cloud: enabling technologies for privacy-preserving multimedia cloud processing. IEEE Signal Process. Mag. **30**(2), 29–41 (2013)
29. Valsesia, D., Coluccia, G., Bianchi, T., Magli, E.: Large-scale image retrieval based on compressed camera identification. IEEE Trans. Multimed. **17**(9), 1439–1449 (2015)
30. Valsesia, D., Coluccia, G., Bianchi, T., Magli, E.: User authentication via PRNU-based physical unclonable functions. IEEE Trans. Inf. Forensics Secur. **12**(8), 1941–1956 (2017)
31. Wang, Q., Zeng, W., Tian, J.: Compressive sensing based secure multiparty privacy preserving framework for collaborative data-mining and signal processing. In: 2014 IEEE International Conference on Multimedia and Expo (ICME), pp. 1–6 (2014)
32. Wang, Q., Zeng, W., Tian, J.: A compressive sensing based secure watermark detection and privacy preserving storage framework. IEEE Trans. Image Process. **23**(3), 1317–1328 (2014)
33. Wu, T., Ruland, C.: Authenticated compressive sensing imaging. In: 2017 International Symposium on Networks, Computers and Communications (ISNCC), pp. 1–6 (2017)
34. Zhou, S., Lafferty, J., Wasserman, L.: Compressed and privacy-sensitive sparse regression. IEEE Trans. Inf. Theory **55**(2), 846–866 (2009)

Chapter 2
Compressed Sensing and Security

Abstract In this chapter we briefly review the Compressed Sensing (CS) framework, discussing the acquisition model, the conditions under which the signal can be recovered, and the main reconstruction algorithms. Then, we show how CS is essentially analogous to a private key cryptosystem if signal acquisition, signal recovery, and sensing matrix generation are interpreted as encryption, decryption, and key generation functions respectively. The basic security properties of this CS cryptosystem under different attack scenarios are discussed according to standard security definitions. This sets the basis for the identification of the attack scenarios that will be analyzed more in depth in Chap. 3. In the second part of this chapter, we introduce the concept of signal embeddings, which can be seen as a generalization of CS measurements. The properties of some of the most common embeddings are briefly reviewed, followed by a discussion on how embeddings can provide privacy-preserving functionalities in particular settings.

Compressed Sensing (CS) [21] is nowadays a popular approach used to simultaneously acquire and compress signals within a single operation. Acquisition and compression can be modeled with a linear operation as

$$\mathbf{y} = \mathbf{\Phi}\mathbf{x} \tag{2.1}$$

where $\mathbf{x} \in \mathbb{R}^n$, $\mathbf{\Phi} \in \mathbb{R}^{m \times n}$, $\mathbf{y} \in \mathbb{R}^m$ and $m \ll n$. The vector $\mathbf{y}$ is a more compact representation of signal $\mathbf{x}$, hence the name "compressed" sensing, and its entries are referred to as measurements. The number of measurements required to represent a signal can be smaller than what is dictated by the Nyquist–Shannon sampling theorem as such theorem only presents a sufficient condition. Signal reconstruction from a set of measurements is possible provided the signal admits a low-dimensional representation. Typically, such low-dimensionality assumption is in the form of sparsity in a transform domain, i.e., when the signal is represented under a suitable basis, only few coefficients are nonzero. Alternatively, a signal is said to be "compressible" if the magnitude of its coefficients decays quickly in some domain, e.g., with a power law. Figure 2.1 shows a graphical depiction of CS.

M. Testa et al., *Compressed Sensing for Privacy-Preserving Data Processing*,
SpringerBriefs in Signal Processing, https://doi.org/10.1007/978-981-13-2279-2_2

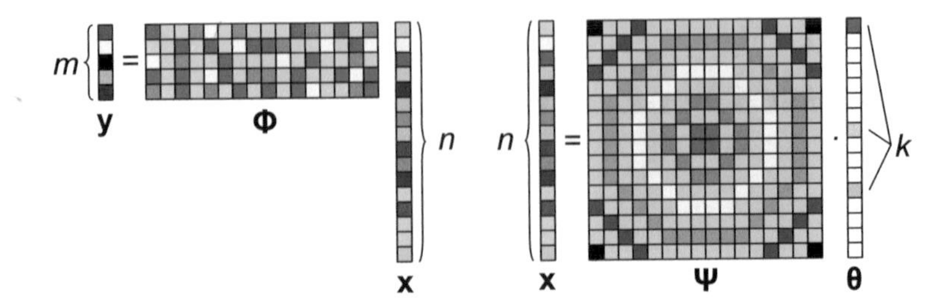

Fig. 2.1 Compressed sensing. A small number of measurements $\mathbf{y}$ of a signal $\mathbf{x}$ is acquired with a sensing matrix $\mathbf{\Phi}$. The signal $\mathbf{x}$ admits a sparse representation under basis $\mathbf{\Psi}$

The breakthrough that started the whole field of CS is that reconstruction of a signal from its measurements is possible by solving a convex optimization problem. Let us consider a signal $\mathbf{x}$ admitting a sparse representation $\mathbf{x} = \mathbf{\Psi}\boldsymbol{\theta}$, with $\|\boldsymbol{\theta}\|_0 = |\{\boldsymbol{\theta}_i \neq 0, i \in [1, n]\}| = k$; this signal is said to be k-sparse if $\mathbf{\Psi}$ is the identity matrix, and k-sparse in a transform domain otherwise. Given a vector of measurements $\mathbf{y}$, the reconstruction of $\mathbf{x}$ is carried out by exploiting the knowledge of the sensing matrix $\mathbf{\Phi}$, basis $\mathbf{\Psi}$, and the fact that $\boldsymbol{\theta}$ is sparse. The most intuitive way to put this knowledge into the form of an optimization problem leads to the following combinatorial problem:

$$\hat{\boldsymbol{\theta}} = \arg\min_{\boldsymbol{\theta}} \|\boldsymbol{\theta}\|_0 \;\text{ s.t. }\; \mathbf{y} = \mathbf{\Phi\Psi}\boldsymbol{\theta} \tag{2.2}$$

i.e., trying to minimize the ℓ_0 pseudonorm, i.e., the number of nonzero entries of a vector. However, solving the above problem is NP-hard. The breakthrough result was realizing that a convex relaxation of this optimization problem, using the ℓ_1 norm instead of the ℓ_0 pseudonorm, is equivalent to the combinatorial optimization problem above under certain conditions. This results in the following optimization problem, which can be solved efficiently:

$$\hat{\boldsymbol{\theta}} = \arg\min_{\boldsymbol{\theta}} \|\boldsymbol{\theta}\|_1 \;\text{ s.t. }\; \mathbf{y} = \mathbf{\Phi\Psi}\boldsymbol{\theta}. \tag{2.3}$$

This is also known as the Basis Pursuit problem [17]. An alternative formulation casts the problem into its unconstrained version, and is typically referred to as Lasso:

$$\hat{\boldsymbol{\theta}} = \arg\min_{\boldsymbol{\theta}} \|\boldsymbol{\theta}\|_1 + \lambda\|\mathbf{y} - \mathbf{\Phi\Psi}\boldsymbol{\theta}\|_2^2.$$

By virtue of Lagrange multipliers, the Lasso and Basis Pursuit formulations are equivalent for a specific choice of λ.

From now on, for the remainder of this book, unless differently specified we will assume that $\mathbf{\Psi}$ is the identity matrix for convenience of explanation, i.e. signal $\mathbf{x}$ is k-sparse. Figure 2.2 shows an intuitive way of explaining why minimizing the ℓ_1 norm leads to sparse solutions for $n = 2$ and $m = 1$. The line represents the space

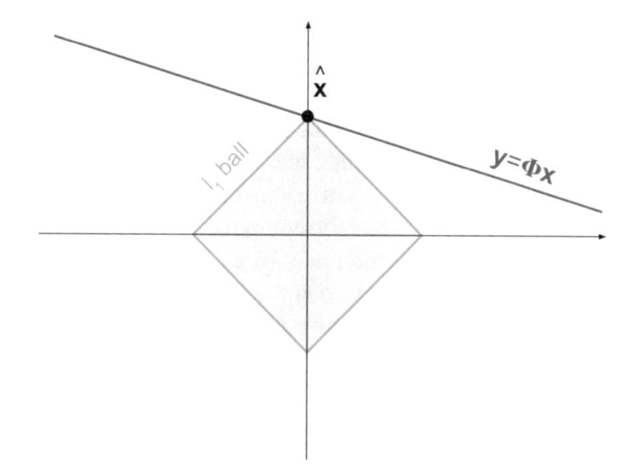

Fig. 2.2 Signal reconstruction via ℓ_1 minimization

of solutions such that $\mathbf{y} = \mathbf{\Phi x}$. One can grow the ℓ_1 ball until it touches the space of solutions, and see that its shape promotes a sparse solution.

Nevertheless, requirements on the structure of both the sensing matrix and the original signal have to be satisfied in order to make the original signal $\mathbf{x}$ the unique solution to the above problem. These requirements can be summarized by the Restricted Isometry Property (RIP) [15].

Definition 2.1 (*Restricted Isometry Property*) A matrix $\mathbf{\Phi}$ satisfies the restricted isometry property (RIP) of order k if there exists a $\delta_k \in (0, 1)$ such that

$$(1 - \delta_k)\|\mathbf{x}\|_2^2 \leq \|\mathbf{\Phi x}\|_2^2 \leq (1 + \delta_k)\|\mathbf{x}\|_2^2$$

for all k-sparse vectors $\mathbf{x}$.

That is, if a matrix satisfies the RIP, then the energy of k-sparse signals is approximately preserved in the compressed domain. The RIP can be used to establish a guarantee on the reconstruction performance of the Basis Pursuit reconstruction algorithm as reported in the following theorem.

Theorem 2.1 (Theorem 1.1 of [12]) *Suppose that $\mathbf{\Phi}$ satisfies the RIP of order $2k$ with $\delta_{2k} < \sqrt{2} - 1$, then the solution $\hat{\mathbf{x}}$ to (2.3) obeys*

$$\|\hat{\mathbf{x}} - \mathbf{x}\|_2 \leq C \frac{\sigma_k(\mathbf{x})_1}{\sqrt{k}}$$

being $\sigma_k(\mathbf{x})_1 = \|\mathbf{x}_k - \mathbf{x}\|_1$ the ℓ_1 norm of the approximation of signal $\mathbf{x}$ with a k-sparse version obtained keeping only the entries with largest magnitude.

Notice that $\sigma_k(\mathbf{x})_1 = 0$ when the signal is exactly k-sparse, so perfect reconstruction can be achieved.

Therefore, the RIP gives us a design criterion to identify the classes of sensing matrices which allow a successful recovery. Several constructions of sensing matrices have been explored in the literature. The most important ones are random matrices with i.i.d. sub-Gaussian entries, i.e. whose tails decay as fast as the ones of the Gaussian distribution. Such matrices satisfy the RIP with high probability for $m > ck \log(n/k)$ for some positive constant c [2]. However, they require the generation of mn random samples and the computation of the full matrix-vector product to obtain the measurements, which may be expensive. Faster solutions have been studied by introducing some structure in the sensing matrix. Examples include sparse random matrices [26], matrices constructed from expander graphs [31], Toeplitz and circulant matrices [40], block-diagonal random matrices [23].

CS allows to achieve significant undersampling factors, i.e. a number of measurements much smaller than the original dimensionality of the signal. However, when considering compression, real-valued measurements must be quantized in order to obtain a representation using a finite rate. It is possible to quantize CS measurements down to 1 bit [30] while still being able to recover the original signal, provided that the reconstruction is "consistent", i.e., the measurements of the reconstructed signal must fall into the original quantization bins. It is important to notice that the naive choice of a uniform scalar quantization of the measurements causes CS-based compression methods to suffer from poor rate-distortion performance. Essentially, the undersampling performed during CS acquisition behaves like oversampling when sparsity is taken into account and its performance falls in line with established results on scalar quantization of oversampled signals [27]. In a nutshell, while increasing the number of quantization levels achieves an exponential reduction in the distortion of the reconstructed signal, increasing the number of measurements only provides a linear reduction in distortion. Recently, novel quantizer designs have been studied to achieve an exponential reduction in distortion [7, 28, 44].

It is worth noting that the literature on CS has seen the development of a multitude of reconstruction algorithms, which can be broadly categorized into three classes: methods based on convex optimization, methods based on greedy algorithms, and methods based on approximate message passing. The class of methods based on optimization stems from the original ℓ_1 minimization formulation in (2.3) [3, 9, 13, 14, 46]. Interior-point optimization methods typically have high computational complexity resulting in slow or prohibitive running time. To speed up the computation, iterative and greedy algorithms have been proposed to perform the optimization. Among the iterative algorithms we mention proximal gradient methods [1, 6, 18, 19, 29, 29, 47]. They are more accurate than greedy algorithms, at the cost of higher computational complexity. As an example, Algorithm 1 reports the pseudocode for the Iterative Soft Thresholding algorithm (ISTA) which is based on proximal gradient descent. Soft thresholding is the proximal mapping operator to the ℓ_1 norm and intuitively promotes sparsity.

Algorithm 1 ISTA

Input: Sensing matrix $\mathbf{\Phi}$, measurements $\mathbf{y}$,
shrinkage operator $\eta_\lambda[\cdot] = \mathrm{sgn}(\cdot)\max(|\cdot| - \lambda, 0)$
Set $\mathbf{x}^{(0)} = \mathbf{0}$, iterate
for $t = 1$ to StopIter **do**
$\mathbf{x}^{(t)} \leftarrow \eta_\lambda[\mathbf{x}^{(t-1)} + \mathbf{\Phi}^T(\mathbf{y} - \mathbf{\Phi}\mathbf{x}^{(t-1)})]$
end for

Greedy algorithms [41, 43], generally, build up an approximation to the solution one step at a time by making locally optimal choices at each step. Examples include Orthogonal Matching Pursuit (OMP) [42], Stagewise OMP (StOMP) [22], Regularized OMP (ROMP) [36], and CoSaMP [35]. Although greedy algorithms are extremely fast, they generally require a larger number of measurements that may not be easy to acquire. Algorithm 2 reports the pseudocode for the OMP algorithm, where $\mathbf{\Phi}_{\mathcal{S}}^{\dagger} = (\mathbf{\Phi}_{\mathcal{S}}^T \mathbf{\Phi}_{\mathcal{S}})^{-1} \mathbf{\Phi}_{\mathcal{S}}^T$ denotes the pseudoinverse of the sensing matrix restricted to the columns in set $\mathcal{S}$. OMP builds the support of the sparse signal by selecting one entry at a time. This entry is chosen as the one that maximizes the inner product between the corresponding column of the sensing matrix and a residual, initialized as the measurements.

Algorithm 2 OMP

Input: Sensing matrix $\mathbf{\Phi}$, measurements $\mathbf{y}$, sparsity k
Set $\mathbf{r}^{(0)} = \mathbf{y}$, $\mathbf{x}^{(0)} = \mathbf{0}$, $\mathcal{S}^{(0)} = \emptyset$, iterate
for $i = 1$ to k **do**
$\mathbf{g}^{(i)} \leftarrow \mathbf{\Phi}^T \mathbf{r}^{(i-1)}$
$j^{(i)} \leftarrow \arg\max_j \frac{|g_j^{(i)}|}{\|\mathbf{\Phi}_j\|_2}$
$\mathcal{S}^{(i)} \leftarrow \mathcal{S}^{(i-1)} \cup j^{(i)}$
$\mathbf{r}^i = \mathbf{y} - \mathbf{\Phi}(\mathbf{\Phi}_{\mathcal{S}^{(i)}}^{\dagger} \mathbf{y})$
end for
$\hat{\mathbf{x}} = \mathbf{\Phi}_{\mathcal{S}^{(i)}}^{\dagger} \mathbf{y}$

Finally, the most recent class of reconstruction algorithm is based on approximate message passing techniques. The most important of these algorithms is Generalized Approximate Message Passing (GAMP) [39]. GAMP takes a Bayesian estimation approach to signal recovery and tackles the computational intractability by reducing a vector-valued estimation problem into a sequence of scalar problems.

2.1 Compressed Sensing as a Cryptosystem

The fact that CS can be efficiently implemented using randomly generated sensing matrices means that CS measurements are naturally equipped with some secrecy notions, that allow us to emply CS as an effective cryptosystem. Indeed, because of

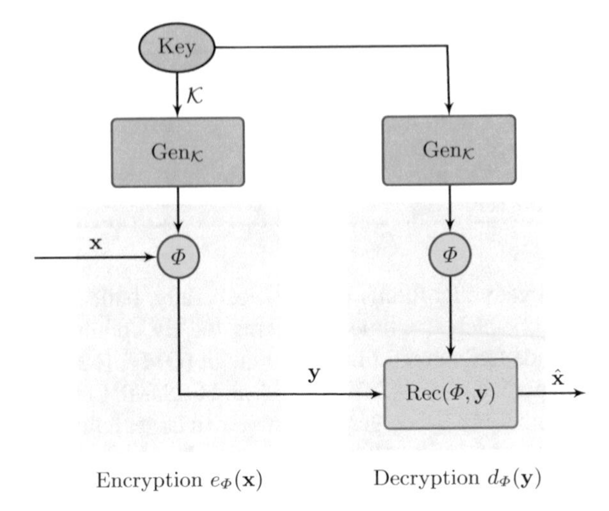

Fig. 2.3 Scheme of a compressive cryptosystem

its structure, the CS acquisition model lends itself to providing confidentiality, since in order to recover a signal given its measurements, the knowledge of the sensing matrix is necessary. Therefore, only those parties who have access to the sensing matrix should be able to recover the original signal. Starting from this consideration, if a party does not have access to the sensing matrix $\mathbf{\Phi}$ but has only access to the measurements $\mathbf{y}$, is it then possible to prove that the original signal cannot be recovered and hence that CS can also provide secrecy? The short answer to this extensively investigated question is yes: the CS acquisition can act as an encryption function. However, if we consider a complete private key cryptosystem, we still need to define a suitable decryption function. Interestingly, CS already provides algorithms which can be used to recover the original signal and thus can be formulated as decryption functions.

Let us formalize these concepts and reconsider CS under a cryptographic perspective. A compressive cryptosystem, as depicted in Fig. 2.3, can be defined as follows. The signal $\mathbf{x}$ is the plaintext, the measurements $\mathbf{y}$ are the ciphertext, and the sensing matrix $\mathbf{\Phi}$ is the secret key. The system is composed of the following functional blocks:

Encryption The encryption $e_{\mathbf{\Phi}}(\mathbf{x})$ is performed through CS acquisition as defined in (2.1), i.e., in a compressive cryptosystem, the encryption function $e_{\mathbf{\Phi}}(\mathbf{x}) = \mathbf{\Phi}\mathbf{x}$ is a simple linear combination of the plaintext with random weights. In our model, we assume that the ciphertext $\mathbf{y} = e_{\mathbf{\Phi}}(\mathbf{x})$ is received without errors. In practical CS applications, it is customary to assume that measurements $\mathbf{y}$ are contaminated by some noise due, e.g., to quantization or channel introduced errors. While this may be an issue concerning the recovery performance of CS, it does not affect the security of the system. As it will become clear in the following, thanks to the data processing inequality any degradation of the measurements will not introduce any advantage for an adversary.

Decryption The decryption $d_{\mathbf{\Phi}}(\mathbf{y})$ corresponds to any CS recovery algorithm that can be used to recover the original signal. The goal of this function is to recover the plaintext given the knowledge of both the secret key and the ciphertext. It can be defined as $d_{\mathbf{\Phi}}(\mathbf{y}) = \text{Rec}(\mathbf{y}, \mathbf{\Phi})$, where $\text{Rec}(\mathbf{y}, \mathbf{\Phi})$ correspond to any suitable CS recovery algorithm, e.g. those described earlier in this section. As introduced at the beginning of this chapter, if some conditions on the number of measurements and on the structure of the sensing matrix are satisfied, then sparse or approximately sparse signals can be recovered with overwhelming probability. Thus, different existing recovery algorithms can be cast into the decryption function $d_{\mathbf{\Phi}}(\mathbf{y})$. Here, it is important to highlight that in general the decryption function is able to *approximately* invert the result of the encryption, i.e., $d_{\mathbf{\Phi}}(e_{\mathbf{\Phi}}(\mathbf{x})) = \hat{\mathbf{x}} \sim \mathbf{x}$, rather than *exactly* as it happens in regular private key cryptosystems.[1] While this does not pose major issues for CS acquisition, since in many scenarios exact recovery of the original signal is not always needed, it has some effect on the security of the CS cryptosystem, since even an approximate decryption may be considered a successful attack.

Key Generation The key of the CS cryptosystem, i.e., the sensing matrix $\mathbf{\Phi}$, is generated at both encryption and decryption side using a common key generation function $\text{Gen}_{\mathcal{K}}$ that relies on a shared secret key $\mathcal{K}$. This solution avoids transmitting a large number of matrix entries to the receiving party. The easiest solution is to design a deterministic key generation function, so that, given $\mathcal{K}$, every plaintext $\mathbf{x}$ is encrypted using the same sensing matrix. However, in the following we will mostly consider a randomized key generation function that gives a different sensing matrix at each sensing step. This strategy will be named one-time sensing (OTS), since it is similar in principle to the well-known one-time pad cryptosystem. As will become clear in the following sections, OTS is essential for guaranteeing the security of the CS cryptosystem. In this book, we will assume that $\text{Gen}_{\mathcal{K}}$ is an ideal function that generates sensing matrices exactly according to a desired distribution. In practice, cryptographic key derivation functions such as SHA-3 (Keccak) [4] can be employed to generate the entries of the secret key. For the interested reader, this particular problem has been addressed in more detail in, e.g., [20, 24, 25].

From the above description, it is evident that a CS cryptosystem is designed according to Kerckhoffs' principle [33]: both encryption and decryption functions are public, the security relies only on the secrecy of the sensing matrix $\mathbf{\Phi}$, or equivalently, the shared secret key $\mathcal{K}$. Nevertheless, the compressive cryptosystem we just defined has some vulnerabilities which have to be addressed by a careful analysis of its components. For this reason, we need to introduce appropriate security definitions and discuss in detail the main assumptions to be satisfied in order to make the cryptosystem resilient to common attack scenarios.

---

[1] Exact reconstruction is possible if $\mathbf{x}$ is k-sparse, as described earlier in this section.

2.1.1 Security Definitions

Even though the secret key is not known, the linear acquisition process of CS may lead to leaks of information through the measurements. The analysis of the information leakage, which is covered in Chap. 3, relies on information theoretic tools and definitions which we briefly recall here. It is important to highlight that the information theoretic approach is stronger than the computational one since it characterizes the amount of information an attacker can have access to. If no sufficient information is available, then even with unbounded computational capabilities the attack cannot succeed.

Given the cryptosystem we defined above, different metrics can be used to characterize its security properties. From an information theoretic perspective, a cryptosystem is said to achieve *perfect secrecy* if

$$\mathbb{P}[\mathbf{y}|\mathbf{x}] = \mathbb{P}[\mathbf{y}]$$

where $\mathbf{y}$ denotes the ciphertext and $\mathbf{x}$ denotes the plaintext. Namely, the posterior probability of the ciphertext given the plaintext is independent of the plaintext. This implies that an attacker cannot be more successful than random guessing the plaintext.

Perfect secrecy is very difficult to achieve in practical cryptosystems, so this definition is usually relaxed by introducing some additional assumptions. If we allow the adversary to have a tiny advantage with respect to random guessing, we say that a cryptosystem achieves *statistical secrecy*. More formally, statistical secrecy can be defined by the following game. Let us consider two messages $\mathbf{x}_1$ and $\mathbf{x}_2$, and randomly encrypt one as $\mathbf{y}$. An adversary observes $\mathbf{y}$ and decides whether this is the encryption of $\mathbf{x}_1$ or $\mathbf{x}_2$. The encryption is said to provide statistical secrecy if

$$\mathbb{P}[\text{success}] = \frac{1}{2} + \epsilon \tag{2.4}$$

where ϵ is negligible with respect to the size of the key. In the following, we will show that several implementations of the CS cryptosystem achieve a weak form of statistical secrecy, in which ϵ does not decrease exponentially with the size of the key.

The above definitions are quite strong, since they rely only on the statistical properties of CS measurements, without making any assumption on the resources of the adversary. However, they usually hold under the ideal assumption that measurements are continuous and sensing matrices are modeled by continuous probability distributions. For practical cases in which both measurements and sensing matrices will be represented using a finite, although very large, set of possible values, another widely used definition is that of *computational secrecy*. In this case, a cryptosystem is said to be computationally secure if the definition in (2.4) holds for any adversary limited to algorithms that run in time polynomial with respect to the size of the key.

2.1.2 Attack Scenarios

In modern cryptography attacks against encryption schemes are usually classified into four types, according to the resources of the adversary:

- *Ciphertext-only attack* (COA): In this scenario the adversary observes only the ciphertext $\mathbf{y}$ (or multiple ciphertexts) and tries to recover the plaintext;
- *Known-plaintext attack* (KPA): In this case the adversary obtains pairs of plaintext/ciphertext encrypted with the same key and tries to decrypt other ciphertexts;
- *Chosen-plaintext attack* (CPA): Here, the adversary can choose arbitrary plaintexts and obtain the corresponding ciphertext, with the aim of decrypting other ciphertexts;
- *Chosen-ciphertext attack* (CCA): In this last scenario, the adversary can obtain the plaintext corresponding to arbitrary ciphertexts, excluding the ciphertexts that are currently under attack.

The linearity of the acquisition process immediately implies that a compressive cryptosystem cannot be secure under either CPA or KPA when the same secret key is re-used at different encryptions. In the case of CPA, if the sensing matrix is kept fixed, by choosing the canonical basis as plaintext, the ciphertext will *exactly* reveal the columns of the secret key $\mathbf{\Phi}$. Differently, in the case of KPA, the secret key cannot be trivially revealed as in the previous scenario; however given n linearly independent plaintext-ciphertext pairs it is possible to solve the system of mn unknowns corresponding to the entries of $\mathbf{\Phi}$. Thus, a KPA attack will also succeed.

Luckily, the OTS strategy can effectively address both attack scenarios. Regenerating the secret key at each encryption will make recovering the sensing matrix unfeasible under both CPA and KPA, since both attack strategies rely on different and subsequent encryptions to be performed with the same key in order to extract its full content. For this reason, in Chap. 3 we will concentrate our attention on OTS cryptosystems. Moreover, if we assume that different sensing matrices are independently generated, KPA or CPA do not provide any advantage over COA, so all the results will be derived under the COA scenario.

It is worth noting that a practical key generation function will unavoidably introduce correlations among sensing matrices, which may be exploited under either KPA or CPA. Although this analysis is out of the scope of this book, the interested reader can find some preliminary results in [11].

Regarding the security of the CS cryptosystem under the COA scenario, it is important to notice that a RIP-satisfying sensing matrix, although being required for signal recovery, is not necessarily the best option for providing secrecy. Being the RIP an ℓ_2-norm preserving property, it is easy to foresee that the energy of the plaintext will be leaked through the measurements. For this reason, it is important to analyze the secrecy related to the linear acquisition model and characterize the amount of information that is leaked through the measurements given a specific choice of sensing matrix. This important problem is addressed in Chap. 3, where the information leakage of different sensing matrices is analyzed.

Table 2.1 Summary of security properties of compressive cryptosystems under different attack scenarios. CS: same sensing matrix is re-used; OTS: sensing matrix re-generated at each encryption. The table indicates whether the system can provide some notion of secrecy or not. OTS is discussed in Chap. 3. Legend: S: statistical secrecy, C: computational secrecy

	COA	KPA	CPA	CCA
CS	C [38]	No	No	No
OTS	S, C [5, 10]	C [11]	C	No

Lastly, we can immediately verify that a CS cryptosystem cannot be secure under CCA, even when using the OTS strategy. Given a targeted encryption $\mathbf{y}$, under CCA the adversary can obtain the decryption $\mathbf{x}'$ of $\mathbf{y}' = \mathbf{y} + \mathbf{n}$, where $\mathbf{n}$ is an arbitrarily small error vector. Due to the linearity of the acquisition process, $\mathbf{x}'$ will be close to $\mathbf{x}$, meaning that the adversary can always obtain a good approximation of the plaintext, or distinguish encryptions of different plaintexts with a success probability close to one. The security properties of compressive cryptosystems under different attack scenarios are summarized in Table 2.1.

2.2 Signal Embeddings

Signal representations seek to capture all the signal information in a compact manner in order to increase a system efficiency in terms of storage or processing requirements. The previous section introduced representations where the primary goal was to reconstruct the original signal from the compact encoding with a low distortion, or, under some conditions, perfectly. However, this is only one facet of the signal representation problem. Oftentimes, we are not directly interested in the signal itself, but rather in the information that can be extracted from it through processing. It is therefore the case that this inference process can be performed more efficiently if the representation of the signal seeks to preserve the information relevant to the processing algorithm rather the the entire signal. A classic example of this is information retrieval, where one has to decide which among many stored signals better matches a query template according to some problem-dependent criterion, e.g., a distance in a metric space. Such problem is not concerned with the actual signals but rather the geometry of the entire set of signals in the form of their pairwise distances. Similarly, many problems in machine learning such as clustering, anomaly detection, regression, classifiers rely on the assumption that the information relevant for the solution of the problem is encoded in the geometry of the signal set and a proper model using the distances between signals achieves the desired solution. Hence, a smart signal representation should seek to provide compact codes in a space with a distance function that is fast to evaluate and that approximately preserves the distances in the original space. This would allow an advantageous trade-off between computational efficiency and performance in terms of the problem-dependent metric.

Embeddings are transformations of a set of signals from a high-dimensional metric space to a low-dimensional space such that the geometry of the set is approximately preserved (or conveniently distorted, in some cases). Distances in the original metric space are replaced by distances measured directly in the low-dimensional space. Thanks to the lower dimensionality and, possibly, a distance function that is faster to compute, inference problems only requiring geometric information about the signal set are greatly accelerated and require less storage.

Definition 2.2 (*Embedding*) A function $f : \mathcal{X} \rightarrow \mathcal{Y}$ is a (g, δ, ϵ)-embedding of metric space $(\mathcal{X}, d_{\mathcal{X}})$ into metric space $(\mathcal{Y}, d_{\mathcal{Y}})$ if, for all $\mathbf{u}, \mathbf{v} \in \mathcal{X}$, it satisfies

$$(1-\delta)g\left(d_{\mathcal{X}}(\mathbf{u}, \mathbf{v})\right) - \varepsilon \leq d_{\mathcal{Y}}\left(f(\mathbf{u}), f(\mathbf{v})\right) \leq (1+\delta)g\left(d_{\mathcal{X}}(\mathbf{u}, \mathbf{v})\right) + \varepsilon.$$

The quantities δ and ϵ are multiplicative and additive ambiguities due to the embedding, respectively. Notice that, in general, distances $d_{\mathcal{Y}}$ in the embedding space may be distorted with respect to original distances $d_{\mathcal{X}}$ due to map $g(\cdot)$.

The most famous result concerning embeddings is the Johnson - Lindenstrauss (JL) lemma [32]. Johnson and Lindestrauss demonstrated that it is possible to use a Lipschitz function f to implement an embedding that preserves Euclidean distances up to a multiplicative distortion.

Lemma 2.1 (Johnson–Lindenstrauss) *Let* $\delta \in (0, 1)$. *For every set* $\mathcal{X}$ *of* $|\mathcal{X}|$ *points in* $\mathbb{R}^n$, *if* m *is a positive integer such that* $m = \mathcal{O}\left(\delta^{-2}\log(|\mathcal{X}|)\right)$, *there exists a Lipschitz mapping* $f : \mathbb{R}^n \rightarrow \mathbb{R}^m$ *such that*

$$(1-\delta)\|\mathbf{u}-\mathbf{v}\|_2^2 \leq \|f(\mathbf{u}) - f(\mathbf{v})\|_2^2 \leq (1+\delta)\|\mathbf{u}-\mathbf{v}\|_2^2 \tag{2.5}$$

for all $\mathbf{u}, \ \mathbf{v} \in \mathcal{X}$.

It is interesting to notice that the number of measurements m required to make the embedding work depends only on the desired fidelity on the Euclidean distances between points, controlled by the distortion parameter δ, and by the number of signals in the set to be embedded. Crucially, there is no dependence on the dimensionality n of input space.

The JL lemma posits the existence of an embedding approximately preserving Euclidean distances. A practical implementation satisfying the Johnson-Lindenstrauss lemma can be obtained by using a random linear mapping (also called random projection)

$$\mathbf{y} = f(\mathbf{x}) = \mathbf{\Phi}\mathbf{x}$$

for $\mathbf{x} \in \mathcal{X} \subset \mathbb{R}^n$ and $\mathbf{y} \in \mathcal{Y} \subset \mathbb{R}^m$ and a random matrix $\mathbf{\Phi}$. It is easy to show that if the entries of $\mathbf{\Phi}$ are drawn independently from specific distributions, then pairwise Euclidean distances between points in $\mathcal{X}$ are equal, within a multiplicative distortion δ, to the distances between their embeddings in $\mathcal{Y}$, with high probability. In particular, if the entries of $\mathbf{\Phi}$ are independent and identically distributed according to a Gaussian, Rademacher, or uniform distribution then Eq. (2.5) holds with probability greater than $1 - c_1 e^{\log|\mathcal{X}| - c_2\delta^2 m}$ for some universal constants c_1 and c_2. Simple proofs for such result can be found [2]. Typically they are composed of two main steps: (i) a concentration of measure result on a pair of signals, (ii) a union bound over all the possible pairs of signals in the set. The former step is concerned with showing that for a fixed pair of signals the distance in the embedding space concentrates around its mean, i.e. deviation from the mean (i.e., the multiplicative distortion of the embedding) vanishes exponentially fast as a function of the distortion and of the number of measurements. The latter step generalizes the previous results to all possible pairs of signals in the set via a union bound argument.

The distance-preservation property of the Johnson–Lindenstrauss lemma closely resembles the Restricted Isometry Property (RIP) introduced in Definition 2.1 in the previous section. However, the key difference is that the JL lemma is concerned with sets with a finite number of points, while the RIP deals with infinite sets such as all the possible k-sparse signals. This makes the RIP a stronger condition with respect to the JL lemma. In fact, it is possible to show that random matrices satisfying the RIP also satisfy the JL lemma [34]. Contrary to the JL lemma, the RIP requires a number of measurements depending on the input dimensionality n. The basic proofs for the RIP closely resemble the ones for the JL lemma. However, after a concentration result is shown for a pair of signal, it must be extended to all the pairs in the infinite set of signals. This is typically done by set covering arguments such as nets of points.

Reducing the dimensionality of the space is only one part of the work to produce compact signal representations. In fact, the embeddings must be quantized in order to reduce the rate for transmission or storage. However, quantization introduces further distortion in the embedding, which can be typically modeled as an additive distortion ε. A well-known example of quantized embedding is sign random projections [16]

$$\mathbf{y} = \mathrm{sign}(\mathbf{\Phi x})$$

where $\mathbf{\Phi}$ is a matrix with independent and identically distributed entries drawn from a Gaussian distribution and the sign function quantizes the random projections to a binary value. Sign random projections are interesting for several reasons. First of all, the distances in the embedding space are computed as Hamming distances, i.e. for embeddings $\mathbf{y} = \mathrm{sign}(\mathbf{\Phi u})$ and $\mathbf{z} = \mathrm{sign}(\mathbf{\Phi v})$

$$d_{\mathcal{Y}} = d_H = \frac{1}{m} \sum_{i=1}^{m} y_i \oplus z_i$$

where

$$y_i \oplus z_i = \begin{cases} 1 & \text{if} \quad y_i \neq z_i \\ 0 & \text{otherwise} \end{cases} .$$

Computing Hamming distances is extremely efficiently as it can be implemented with bitwise XOR operations. Then, sign random projections can be shown to preserve the angle between signals in the original space rather than the Euclidean distance, i.e.,

$$(1 - \delta) d_{\angle}(\mathbf{u}, \mathbf{v}) \leq d_H(\mathrm{sign}(\mathbf{\Phi u}), \mathrm{sign}(\mathbf{\Phi v})) \leq (1 + \delta) d_{\angle}(\mathbf{u}, \mathbf{v})$$

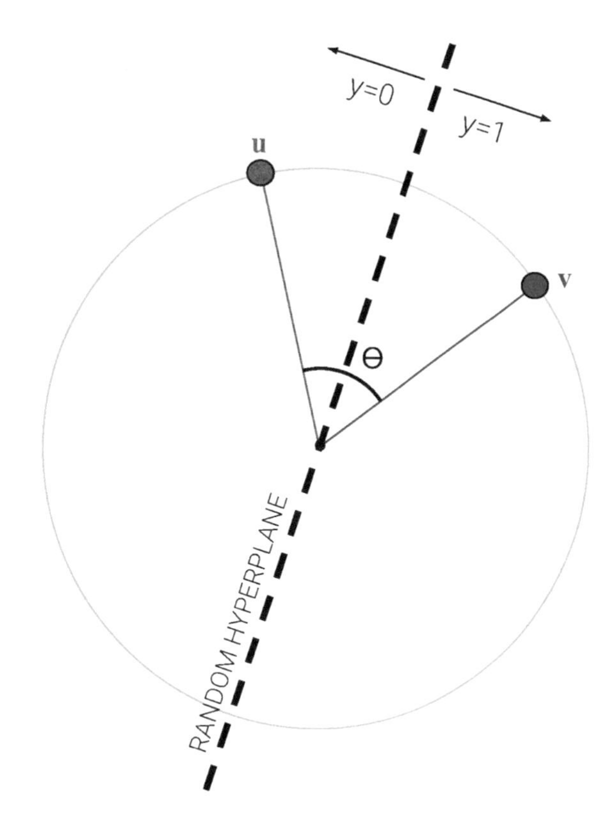

Fig. 2.4 Sign random projections

being

$$d_{\angle}(\mathbf{u}, \mathbf{v}) = \frac{1}{\pi}\arccos\left(\frac{\mathbf{u}^T\mathbf{v}}{\|\mathbf{u}\|\|\mathbf{v}\|}\right)$$

the angular distance between the original signals. This property can be intuitively understood by observing sign random projections from a geometrical standpoint, as depicted in Fig. 2.4. Measurements y_i and z_i are determined by selecting a random hyperplane (a row of matrix $\mathbf{\Phi}$ fixes its normal direction) and if the original signals fall in the same half-space created by the hyperplane then the two measurements will coincide. Since the entries of $\mathbf{\Phi}$ are i.i.d. Gaussians, the normal of the hyperplane is spherically uniform and therefore only the angle between the signals determines the outcome of the measurement.

Preserving angular distances can also be achieved by means of quantized phase embeddings [8]. They can be seen as a generalization of sign random projections, where the phase of a complex-valued random projection is measured instead of the sign. The phase can be further quantized with a uniform scalar quantizer. Therefore, the model of quantized phase embeddings is

$$\mathbf{y} = Q\left(\angle\left(\mathbf{\Phi}\mathbf{x}\right)\right)$$

where $\angle$ is the principal phase operator, $\boldsymbol{\Phi}$ a random matrix with i.i.d. complex Gaussian entries, and Q a uniform scalar quantizer.

Finally, it is worth mentioning embeddings that are adaptive [37, 45], in the sense that provide reduced distortion for signals belonging to a specific class or similar to a reference. A typical approach is to use a projection matrix $\boldsymbol{\Phi}$ optimized from training data according to some criterion, followed by binary quantization. In [45], a binary-quantized adaptive embedding is constructed by selecting the positions of the largest entries among the random projections of a reference signal before quantization. When other signals are embedded the binary values in the corresponding positions are selected. Using a Gaussian sensing matrix to compute the random projections of a reference signal, essentially amounts to drawing some directions uniformly at random on a sphere and projecting the signal onto such directions. The choice of the largest random projections in the adaptive embedding preferentially selects the directions which are best aligned with the reference signal from the ones drawn uniformly at random. Let us call the reference signal $\mathbf{u} \in \mathbb{R}^n$; a number m_{pool} of random projections is computed with a sensing matrix made of i.i.d. Gaussian entries $\boldsymbol{\Phi}_{ij} \sim \mathcal{N}(0, \sigma^2)$ as $\mathbf{y} = \boldsymbol{\Phi}\mathbf{u}$. Then, the m entries with largest absolute value are identified and their locations stored in vector $\mathbf{l}$. The final binary embedding of the reference signal $\mathbf{u}$ is therefore:

$$\mathbf{p} = \text{sign}\left(\boldsymbol{\Phi}_{\mathbf{l}}\mathbf{u}\right)$$

being $\boldsymbol{\Phi}_{\mathbf{l}}$ the restriction of $\boldsymbol{\Phi}$ to the rows identified by $\mathbf{l}$. When computing the binary embedding of a new signal, the same locations $\mathbf{l}$ are used:

$$\mathbf{q} = \text{sign}\left(\boldsymbol{\Phi}_{\mathbf{l}}\mathbf{v}\right).$$

It can be shown that the Hamming distance between the binary-quantized adaptive projections of the reference signal and the binary-quantized adaptive projections of another signal concentrates around the following value:

$$\mathbb{E}\left[d_H(\mathbf{p}, \mathbf{q})\right] = \frac{1}{m}\sum_{i=1}^{m} p_i$$

$$p_i = \frac{1}{2} + \frac{1}{2}\text{erf}\left(-y_{l_i}\frac{\mathbf{u}^T\mathbf{v}}{\sqrt{2}\sigma\|\mathbf{u}\|\sqrt{\|\mathbf{u}\|^2\|\mathbf{v}\|^2 - (\mathbf{u}^T\mathbf{v})^2}}\right)$$

being y_{l_i} the entries of the random projections of the reference signal before quantization. Figure 2.5 shows the behavior of the Hamming distance as a function of the inner product between a unit-norm reference signal and another unit-norm one with respect to sign random projections.

The compact representation of signals provided by embeddings is obviously useful in terms of providing a more efficient processing with lower computational or storage requirements or reduced bandwidth for transmission. Perhaps less intuitively,

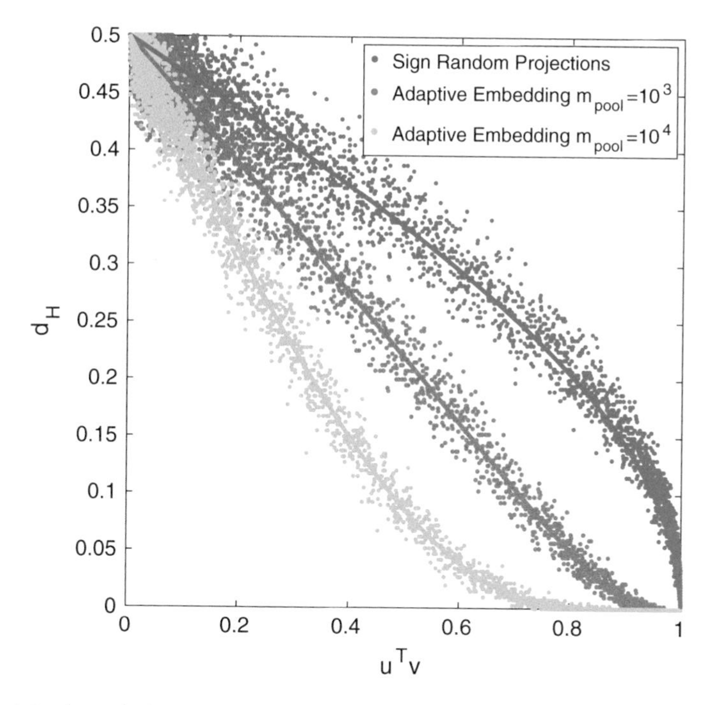

Fig. 2.5 Adaptive embedding and sign random projections. $m = 400$

embeddings can also provide privacy-preserving functionalities, whereby only a limited amount of information is maintained and thus potentially exposed by the embedding. In particular, embeddings can be seen as privacy-preserving from two main points of view: hiding the signal, or hiding the geometry of the signal set.

Hiding the signal. Certain applications where the signal of interest may contain sensitive information (e.g., medical records, etc.) could greatly benefit from the capability of processing the signal without revealing its content. As an example, let us think of a medical image to be used to infer whether a patient has a particular disease. The image could be transmitted to and processed at a remote system, e.g. a cloud service, where its sensitive content would be at risk of being exposed to attackers. Embeddings could be beneficial in such scenario since they allow processing without revealing the content of the signal. Section 2.1 discussed how it is possible to recover a signal from its random projections if it satisfies certain low-dimensionality assumptions. However, signal recovery crucially needs the sensing matrix to be known. It is therefore clear that if the processing algorithm only requires the distances among signals in a set, embeddings could provide most of the information for the algorithm without revealing the signal itself.

Hiding the geometry. Certain embedding constructions are able to carry useful information about a pair of signals only if their distance is below a threshold. In essence, a third party that is only able to observe the embeddings, would not be able to tell the relationship between two signals if their original distance was greater than the threshold. This is due to the geometry of the signal set being distorted in such a way that the embeddings of those signals appear to be all at the same distance relative to each other.

We are going to explore those two perspectives more in detail in Chap. 4.

References

1. Ba, D., Babadi, B., Purdon, P.L., Brown, E.N.: Convergence and stability of iteratively reweighted least squares algorithms. IEEE Trans. Signal Process. **62**(1), 183–195 (2014)
2. Baraniuk, R., Davenport, M., DeVore, R., Wakin, M.: A simple proof of the restricted isometry property for random matrices. Constr. Approx. **28**(3), 253–263 (2008)
3. van den Berg, E., Friedlander, M.P.: SPGL1: A solver for large-scale sparse reconstruction (2007)
4. Bertoni, G., Daemen, J., Peeters, M., Van Assche, G.: The keccak sha-3 submission. Submission to NIST (Round 3) **6**(7), 16 (2011)
5. Bianchi, T., Bioglio, V., Magli, E.: Analysis of one-time random projections for privacy preserving compressed sensing. IEEE Trans. Inf. Forensics Secur. **11**(2), 313–327 (2016)
6. Blumensath, T., Davies, M.E.: Iterative hard thresholding for compressed sensing. Appl. Comput. Harmon. Anal. **27**(3), 265–274 (2009)
7. Boufounos, P.T.: Universal rate-efficient scalar quantization. IEEE Trans. Inf. Theory **58**(3), 1861–1872 (2012)
8. Boufounos, P.T.: Angle-preserving quantized phase embeddings. Wavelets and Sparsity XV, vol. 8858, p. 88581C. International Society for Optics and Photonics, Bellingham (2013)
9. Boyd, S., Parikh, N., Chu, E., Peleato, B., Eckstein, J., et al.: Distributed optimization and statistical learning via the alternating direction method of multipliers. Found. Trends®. Mach. Learn. **3**(1), 1–122 (2011)
10. Cambareri, V., Mangia, M., Pareschi, F., Rovatti, R., Setti, G.: Low-complexity multiclass encryption by compressed sensing. IEEE Trans. Signal Process. **63**(9), 2183–2195 (2015)
11. Cambareri, V., Mangia, M., Pareschi, F., Rovatti, R., Setti, G.: On known-plaintext attacks to a compressed sensing-based encryption: a quantitative analysis. IEEE Trans. Inf. Forensics Secur. **10**(10), 2182–2195 (2015)
12. Candès, E.J.: The restricted isometry property and its implications for compressed sensing. Comptes Rendus Mathematique **346**(9), 589–592 (2008)
13. Candès, E.J., Romberg, J.: L1-magic: recovery of sparse signals via convex programming (2005)
14. Candès, E.J., Romberg, J.K., Tao, T.: Stable signal recovery from incomplete and inaccurate measurements. Commun. Pure Appl. Math. **59**(8), 1207–1223 (2006)
15. Candès, E.J., Tao, T.: Decoding by linear programming. IEEE Trans. Inf. Theory **51**(12), 4203–4215 (2005)
16. Charikar, M.S.: Similarity estimation techniques from rounding algorithms. In: Proceedings of the Thiry-Fourth Annual ACM Symposium on Theory of Computing, STOC'02, pp. 380–388. ACM, New York (2002)
17. Chen, S.S., Donoho, D.L., Saunders, M.A.: Atomic decomposition by basis pursuit. SIAM Rev. **43**(1), 129–159 (2001)
18. Daubechies, I., Defrise, M., De Mol, C.: An iterative thresholding algorithm for linear inverse problems with a sparsity constraint. Commun. Pure Appl. Math. **57**(11), 1413–1457 (2004)
19. Daubechies, I., Fornasier, M., Loris, I.: Accelerated projected gradient method for linear inverse problems with sparsity constraints. J. Fourier Anal. Appl. **14**(5–6), 764–792 (2008)

20. Djeujo, R.A., Ruland, C.: Secure matrix generation for compressive sensing embedded cryptography. In: 2016 IEEE 7th Annual Information Technology, Electronics and Mobile Communication Conference (IEMCON), pp. 1–8 (2016)
21. Donoho, D.L.: Compressed sensing. IEEE Trans. Inf. Theory **52**(4), 1289–1306 (2006)
22. Donoho, D.L., Tsaig, Y., Drori, I., Starck, J.L.: Sparse solution of underdetermined linear equations by stagewise orthogonal matching pursuit, submitted to. IEEE Transaction Information Theory, Citeseer (2006)
23. Eftekhari, A., Yap, H.L., Rozell, C.J., Wakin, M.B.: The restricted isometry property for random block diagonal matrices. Appl. Comput. Harmon. Anal. **38**(1), 1–31 (2015)
24. Fay, R.: Introducing the counter mode of operation to compressed sensing based encryption. Inf. Process. Lett. **116**(4), 279–283 (2016)
25. Fay, R., Ruland, C.: Compressive sensing encryption modes and their security. In: 2016 11th International Conference for Internet Technology and Secured Transactions (ICITST), pp. 119–126. IEEE (2016)
26. Gilbert, A., Indyk, P.: Sparse recovery using sparse matrices. Proc. IEEE **98**(6), 937–947 (2010)
27. Goyal, V.K., Vetterli, M., Thao, N.T.: Quantized overcomplete expansions in r^n: analysis, synthesis, and algorithms. IEEE Trans. Inf. Theory **44**(1), 16–31 (1998)
28. Güntürk, C.S., Lammers, M., Powell, A.M., Saab, R., Yılmaz, Ö.: Sobolev duals for random frames and σ δ quantization of compressed sensing measurements. Found. Comput. Math. **13**(1), 1–36 (2012)
29. Huebner, E., Tichatschke, R.: Relaxed proximal point algorithms for variational inequalities with multi-valued operators. Optim. Methods Softw. **23**(6), 847–877 (2008)
30. Jacques, L., Laska, J.N., Boufounos, P.T., Baraniuk, R.G.: Robust 1-bit compressive sensing via binary stable embeddings of sparse vectors. IEEE Trans. Inf. Theory **59**(4), 2082–2102 (2013)
31. Jafarpour, S., Xu, W., Hassibi, B., Calderbank, R.: Efficient and robust compressed sensing using optimized expander graphs. IEEE Trans. Inf. Theory **55**(9), 4299–4308 (2009)
32. Johnson, W.B., Lindenstrauss, J.: Extensions of Lipschitz mappings into a Hilbert space. Contemp. Math. **26** (1984)
33. Katz, J., Lindell, Y.: Introduction to Modern Cryptography. Chapman and Hall/CRC Cryptography and Network Security Series. Chapman & Hall/CRC, London (2007)
34. Krahmer, F., Ward, R.: New and Improved Johnson-Lindenstrauss embeddings via the restricted isometry property. SIAM J. Math. Anal. **43**(3), 1269–1281 (2011)
35. Needell, D., Tropp, J.A.: Cosamp: iterative signal recovery from incomplete and inaccurate samples. Commun. ACM **53**(12), 93–100 (2010)
36. Needell, D., Vershynin, R.: Signal recovery from incomplete and inaccurate measurements via regularized orthogonal matching pursuit. IEEE J. Sel. Top. Signal Process. **4**(2), 310–316 (2010)
37. Norouzi, M., Fleet, D.J., Salakhutdinov, R.R.: Hamming distance metric learning. In: Pereira, F., Burges, C.J.C., Bottou, L., Weinberger, K.Q. (eds.) Advances in Neural Information Processing Systems, pp. 1061–1069. Curran Associates, USA (2012)
38. Rachlin, Y., Baron, D.: The secrecy of compressed sensing measurements. In: 2008 46th Annual Allerton Conference on Communication, Control, and Computing, pp. 813–817. IEEE (2008)
39. Rangan, S.: Generalized approximate message passing for estimation with random linear mixing. In: 2011 IEEE International Symposium on Information Theory Proceedings (ISIT), pp. 2168–2172. IEEE (2011)
40. Rauhut, H.: Circulant and Toeplitz matrices in compressed sensing. In: SPARS'09 - Signal Processing with Adaptive Sparse Structured Representations (2009)
41. Tropp, J.A.: Just relax: convex programming methods for identifying sparse signals in noise. IEEE Trans. Inf. Theory **52**(3), 1030–1051 (2006)
42. Tropp, J.A., Gilbert, A.C.: Signal recovery from random measurements via orthogonal matching pursuit. IEEE Trans. Inf. Theory **53**(12), 4655–4666 (2007)
43. Tropp, J.A., Gilbert, A.C., Strauss, M.J.: Algorithms for simultaneous sparse approximation. Part i: greedy pursuit. Signal Process. **86**(3), 572–588 (2006)

44. Valsesia, D., Boufounos, P.T.: Universal encoding of multispectral images. In: 2016 IEEE International Conference on Acoustics, Speech and Signal Processing (ICASSP), pp. 4453–4457 (2016)
45. Valsesia, D., Magli, E.: Binary adaptive embeddings from order statistics of random projections. IEEE Signal Process. Lett. **24**(1), 111–115 (2017)
46. Van Den Berg, E., Friedlander, M.P.: Probing the Pareto frontier for basis pursuit solutions. SIAM J. Sci. Comput. **31**(2), 890–912 (2008)
47. Wang, M., Xu, W., Tang, A.: On the performance of sparse recovery via ℓ_p-minimization ($0 \leq p \leq 1$). IEEE Trans. Inf. Theory **57**(11), 7255–7278 (2011)

Chapter 3
Compressed Sensing as a Cryptosystem

Abstract This chapter presents the most relevant results on Compressed Sensing (CS) used as a cryptosystem. First, we analyze the statistical properties of CS measurements, showing that they always convey at least the energy of the sensed signal. Then, we discuss the secrecy achievable by different sensing matrix constructions. For sensing matrices made of Gaussian i.i.d. entries, we have the highest secrecy guarantees, where only the energy of the signal can be revealed. This particular case is analyzed by introducing a secrecy metric that depends on the ability to estimate the signal energy by an adversary who observes only the signal measurements. The secrecy achievable by generic sensing matrices is analyzed by introducing a distinguishability metric inspired by the standard statistical secrecy definition used in cryptography. Results are provided for matrices made of i.i.d entries with generic distributions and circulant matrices. At the end of the chapter, we discuss several issues connected with the practical implementation of a CS cryptosystem, including sensing matrix generation and quantization of sensing matrix entries.

The previous chapter gave the reader a flavour of how CS can be cast into a effective cryptosystem. As a result, all the applications which make use of CS can also provide some kind of secrecy, with little or no added cost. This is extremely advantageous for the wide range of low-complexity devices which may acquire sensitive data and might not be able to cope with standard encryption schemes such as AES [23]. Nevertheless, as it has become immediately clear that CS could provide secrecy [22], it has also become evident that the secrecy provided by a CS cryptosystem is, in general, a weak notion of secrecy.

In this chapter we will guide the reader through a detailed analysis of the secrecy of CS cryptosystems based on different kinds of sensing matrices. As a matter of fact, the literature contains several results on the recovery performance when different sensing matrix classes are employed. Hence, it is of paramount importance to characterize their behavior also from a secrecy perspective.

In more detail, our analysis will start in Sect. 3.1 with the characterization of the security properties of the measurements which are inherent to the linear acquisition process. With these tools in hand, we will move to the specific case of sensing matrices made of i.i.d. Gaussian entries in Sect. 3.2.1. In this regard, we will start

M. Testa et al., *Compressed Sensing for Privacy-Preserving Data Processing*,
SpringerBriefs in Signal Processing, https://doi.org/10.1007/978-981-13-2279-2_3

with a characterization of the information leakage, and discuss a suitable metric which can be used to quantify the information leakage. We will also briefly cover the behavior of generic sensing matrices which, in the asymptotic setting, can be assimilated to that of Gaussian sensing matrices.

In Sect. 3.3 we will address in detail the information leakage of arbitrary sensing matrices. After defining a suitable secrecy metric, we will consider the case of generic sensing matrices, as well as that of Gaussian but structured, specifically, circulant sensing matrices. We will also briefly address the case of sensing signals with different energies.

To conclude, in Sect. 3.4, we will move to a more practical understanding of a CS cryptosystem. More specifically, we will give an overview of the issues and related solutions regarding the implementation of a practical CS cryptosystem.

3.1 Statistical Properties of Measurements

In this section, we discuss the statistical properties of the measurements from the secrecy point of view. In more detail, we analyze *what* information is leaked through the measurements because of the linear acquisition model of CS. A more detailed analysis aimed at *quantifying* the information leakage is carried out in Sects. 3.2.1 and 3.3. Let us start with two important definitions which will be used in the following.

Definition 3.1 (*Mutual Information*) The mutual information between two continuous random variables X and Y, denoted as $I(X;Y)$ is defined as

$$I(X;Y) = \int_X \int_Y \mathbb{P}(x, y) \log \frac{\mathbb{P}(x, y)}{\mathbb{P}(x)\mathbb{P}(y)} dxdy.$$

This quantity is an important metric for the characterization of the information leakage. Intuitively, it gives a measure of how much the uncertainty about X (Y) is reduced given the knowledge of Y (X). In a private key cryptosystem, we ideally want that the knowledge of the plaintext X should not decrease the uncertainty about the ciphertext Y. In other words we desire the ciphertext to be statistically independent of the plaintext. The following definition formalizes this concept.

Definition 3.2 (*Perfect secrecy*) Let us call the set of possible plaintexts $\mathcal{P}$, the set of ciphertexts $\mathcal{C}$ and a key $\mathcal{K}$. A private key cryptosystem is a pair of functions $e_{\mathcal{K}} : \mathcal{P} \to \mathcal{C}$, $d_{\mathcal{K}} : \mathcal{C} \to \mathcal{P}$ such that, given a plain text $\mathbf{x} \in \mathcal{P}$, and a ciphertext $\mathbf{y} \in \mathcal{C}$, we have that $d_{\mathcal{K}}(e_{\mathcal{K}}(\mathbf{x})) = \mathbf{x}$ and that it is unfeasible, without knowing the key $\mathcal{K}$, to determine $\mathbf{x}$ such that $e_{\mathcal{K}}(\mathbf{x}) = \mathbf{y}$. A cryptosystem is said to be perfectly secure [25] if the posterior probability of the ciphertext given the plaintext $\mathbf{x}$ is independent of $\mathbf{x}$, i.e., if

$$\mathbb{P}(\mathbf{y}|\mathbf{x}) = \mathbb{P}(\mathbf{y}).$$

For a perfectly secure cryptosystem it follows that $I(\mathbf{y}; \mathbf{x}) = 0$, namely the knowledge of the plaintext $\mathbf{x}$ does not decrease the uncertainty about the ciphertext $\mathbf{y}$. However, given the linear acquisition model of CS, in general it is not possible to achieve perfect secrecy. As proved in [22], for a CS cryptosystem it always holds $I(\mathbf{x}; \mathbf{y}) > 0$. Intuitively, because of the linearity of the acquisition model and since CS requires the sensing matrix to approximately preserve the ℓ_2 norms, $\mathbf{x}$ and $\mathbf{y}$ are statistically dependent. As a simple example, just consider the case $\mathbf{x} = 0$: this immediately implies $\mathbf{y} = 0$, hence $\mathbb{P}(\mathbf{y}|\mathbf{x} = 0) \neq \mathbb{P}(\mathbf{y})$ and $I(\mathbf{x}; \mathbf{y})$ cannot be equal to zero. Nonetheless, as shown in the following, in the case of Gaussian sensing matrices it is possible to achieve perfect secrecy in some particular settings.

By recalling the OTS security model defined in Chap. 2, we start the analysis with a result that holds for generic sensing matrices and that specifies what can be inferred by observing only the direction of the vector of measurements $\mathbf{y}$.

Proposition 3.1 *Let us define the spherical angle of* $\mathbf{x}$ *and* $\mathbf{y}$ *as* $\mathbf{u_x} = \mathbf{x}/\sqrt{\mathcal{E}_\mathbf{x}}$ *and* $\mathbf{u_y} = \mathbf{y}/\sqrt{\mathcal{E}_\mathbf{y}}$, *where* $\mathcal{E}_\mathbf{x} = ||\mathbf{x}||_2^2$ *and* $\mathcal{E}_\mathbf{y} = ||\mathbf{y}||_2^2$ *denote the energy of the plaintext and the ciphertext, respectively. Then, if* $\mathcal{E}_\mathbf{y} > 0$, *generic OTS measurements satisfy* $I(\mathbf{x}; \mathbf{u_y}) = I(\mathbf{u_x}; \mathbf{u_y})$.

Proof This easily follows from the fact that if we consider the equalities $\mathbf{y} = \mathbf{\Phi x} = \sqrt{\mathcal{E}_\mathbf{x}} \cdot \mathbf{\Phi u_x}$ and $\mathcal{E}_\mathbf{y} = \mathbf{y}^T\mathbf{y} = \mathcal{E}_\mathbf{x} \cdot \mathbf{u_x}^T \mathbf{\Phi}^T \mathbf{\Phi u_x}$, then we have that

$$\mathbf{u_y} = \mathbf{y}/\sqrt{\mathcal{E}_\mathbf{y}} = (\mathbf{u_x}^T \mathbf{\Phi}^T \mathbf{\Phi u_x})^{-1/2} \mathbf{\Phi u_x} \tag{3.1}$$

which in turn implies $\mathbb{P}(\mathbf{u_y}|\mathbf{x}) = \mathbb{P}(\mathbf{u_y}|\mathbf{u_x})$. The statement of the theorem can be proved by using the following information equalities

$$\begin{aligned} I(\mathbf{x}; \mathbf{u_y}) =& I(\mathbf{x}, \mathbf{u_x}; \mathbf{u_y}) \\ =& I(\mathbf{u_x}; \mathbf{u_y}) + I(\mathbf{x}; \mathbf{u_y}|\mathbf{u_x}) \\ =& I(\mathbf{u_x}; \mathbf{u_y}). \end{aligned} \tag{3.2}$$

since $\mathbb{P}(\mathbf{u_y}|\mathbf{x}) = \mathbb{P}(\mathbf{u_y}|\mathbf{u_x})$ implies $I(\mathbf{x}; \mathbf{u_y}|\mathbf{u_x}) = 0$. □

Thus, in the case of generic sensing matrices, the information which is leaked through the spherical angle of the measurements is nothing more than the spherical angle of $\mathbf{x}$. In addition, $\mathbf{u_y}$ is a sufficient statistic to estimate $\mathbf{u_x}$.

One may wonder whether a similar result holds for the energy of the measurements. For generic matrices, this is not the case, as can be easily checked by considering that $\mathcal{E}_\mathbf{y} = \mathcal{E}_\mathbf{x} \cdot \mathbf{u_x}^T \mathbf{\Phi}^T \mathbf{\Phi u_x}$, i.e., the energy of the measurements depends on both the energy of the signal and its spherical angle. However, we have two important results that hold in the case of Gaussian sensing matrices which we will refer to as Gaussian One Time Sensing (G-OTS) model. We assume $\mathbf{x}$ being a random vector with an arbitrary probability distribution and $\mathbf{\Phi}$ being made of i.i.d. zero-mean Gaussian entries. Then, as shown in [4], the mutual information between $\mathbf{x}$ and $\mathbf{y}$ $I(\mathbf{x}; \mathbf{y})$ does not depend on $\mathbf{x}$ but solely on its energy $\mathcal{E}_\mathbf{x}$. More specifically we have

Proposition 3.2 *If the entries of* $\mathbf{\Phi}$ *are i.i.d. zero-mean Gaussian, then* $I(\mathbf{x};\mathbf{y}) = I(\mathcal{E}_{\mathbf{x}};\mathbf{y})$.

Proof Let us consider the probability distribution function $\mathbb{P}(\mathbf{y}|\mathbf{x})$ for a given $\mathbf{x}$. Since $\mathbf{\Phi}_{i,j}$ are Gaussian, we have that $\mathbb{P}(\mathbf{y}|\mathbf{x})$ is a multivariate Gaussian distribution with mean $\mu_{\mathbf{y}|\mathbf{x}}$ and covariance matrix $\mathbf{C}_{\mathbf{y}|\mathbf{x}}$. It is immediate to find $\mu_{\mathbf{y}|\mathbf{x}} = E[\mathbf{y}|\mathbf{x}] = E[\mathbf{\Phi}]\mathbf{x} = 0$, whereas by rewriting $\mathbf{y} = (\mathbf{I}_m \otimes \mathbf{x}^T)(\mathbf{\Phi}^T)$, where $(\mathbf{\Phi}^T)$ vectorizes matrix $\mathbf{\Phi}^T$ by stacking its columns, we have

$$\begin{aligned} E[\mathbf{y}\cdot\mathbf{y}^T|\mathbf{x}] =&(\mathbf{I}_m \otimes \mathbf{x}^T)E[(\mathbf{\Phi}^T)(\mathbf{\Phi}^T)^T](\mathbf{I}_m \otimes \mathbf{x}) \\ =&\sigma_{\mathbf{\Phi}}^2(\mathbf{I}_m \otimes \mathbf{x}^T)(\mathbf{I}_m \otimes \mathbf{x}) \\ =&\sigma_{\mathbf{\Phi}}^2\mathbf{x}^T\mathbf{x}\mathbf{I}_m \\ =&\sigma_{\mathbf{\Phi}}^2\mathcal{E}_{\mathbf{x}}\mathbf{I}_m \end{aligned} \tag{3.3}$$

where m is the number of measurements, $\mathbf{I}_m$ denotes an $m \times m$ identity matrix, and we assume that $\mathbf{\Phi}_{i,j}$ have variance $\sigma_{\mathbf{\Phi}}^2$. From the above results, it follows that $\mathbb{P}(\mathbf{y}|\mathbf{x})$ depends only from $\mathcal{E}_{\mathbf{x}}$, i.e. $\mathbb{P}(\mathbf{y}|\mathbf{x}) = \mathbb{P}(\mathbf{y}|\mathcal{E}_{\mathbf{x}})$. The proof then follows from the following chain of mutual information equalities [10]

$$\begin{aligned} I(\mathbf{x};\mathbf{y}) =&I(\mathbf{x},\mathcal{E}_{\mathbf{x}};\mathbf{y}) \\ =&I(\mathcal{E}_{\mathbf{x}};\mathbf{y}) + I(\mathbf{x};\mathbf{y}|\mathcal{E}_{\mathbf{x}}) \\ =&I(\mathcal{E}_{\mathbf{x}};\mathbf{y}). \end{aligned} \tag{3.4}$$

since $\mathbb{P}(\mathbf{y}|\mathbf{x}) = \mathbb{P}(\mathbf{y}|\mathcal{E}_{\mathbf{x}})$ implies $I(\mathbf{x};\mathbf{y}|\mathcal{E}_{\mathbf{x}}) = 0$. □

This means that, when the sensing matrix is made of i.i.d. Gaussian entries, the measurements do not reveal anything more about $\mathbf{x}$ than its energy and the information which can be inferred given the knowledge of $\mathcal{E}_{\mathbf{x}}$. This is an interesting result which allows us to say that G-OTS measurements can achieve *spherical secrecy*. This weaker notion of secrecy is related to the fact that the measurement do not directly expose the spherical angle of the original signal. Moreover, if the energy of $\mathbf{x}$ is independent of its spherical angle, it turns out that the measurements are also independent of the spherical angle of $\mathbf{x}$.

An important consequence of the above results is that G-OTS measurements do not reveal information about signals which lie in a class of equal energy signals. In other words, perfect secrecy can be achieved for those signals which belong to a class of constant energy signals.

Proposition 3.3 *If* $\mathbf{x} \in \mathcal{S}_\beta$, *where* $\mathcal{S}_\beta = \{\mathbf{x}|\mathcal{E}_{\mathbf{x}} = \beta > 0\}$, *then G-OTS measurements are perfectly secure.*

Proof If $\mathcal{E}_{\mathbf{x}}$ is an a priori known constant, then $\mathbb{P}(\mathbf{y}|\mathcal{E}_{\mathbf{x}}) = \mathbb{P}(\mathbf{y})$. Thus, since for G-OTS we have $\mathbb{P}(\mathbf{y}|\mathcal{E}_{\mathbf{x}}) = \mathbb{P}(\mathbf{y}|\mathbf{x})$, it follows that $\mathbb{P}(\mathbf{y}|\mathbf{x}) = \mathbb{P}(\mathbf{y})$. □

Despite the interesting result we showed above, in practice most signals of interest do not belong to classes of equal energy signals. Nevertheless, we can exploit the above result to show that a normalized version of G-OTS measurements achieves perfect secrecy irrespective of the signal distribution. First, we need to show that the energy of $\mathbf{y}$ is a sufficient statistics for estimating $\mathcal{E}_{\mathbf{x}}$.

Proposition 3.4 *Let $\mathcal{E}_{\mathbf{y}} = ||\mathbf{y}||_2$ be the energy of the measurements vector $\mathbf{y}$, and $\mathbf{\Phi}$ be composed of i.i.d. zero-mean Gaussian variables, then we have that $I(\mathcal{E}_{\mathbf{x}}; \mathbf{y}) = I(\mathcal{E}_{\mathbf{x}}; \mathcal{E}_{\mathbf{y}})$.*

Proof For a given $\mathcal{E}_{\mathbf{x}}$, $\mathbf{y}$ is distributed as a multivariate Gaussian with diagonal covariance matrix, hence we have $\mathbb{P}(\mathbf{y}|\mathcal{E}_{\mathbf{x}}) = f(\mathbf{y}^T\mathbf{y}, \mathcal{E}_{\mathbf{x}}) = f(\mathcal{E}_{\mathbf{y}}, \mathcal{E}_{\mathbf{x}})$. Moreover $\mathbb{P}(\mathbf{y}) = \int f(\mathcal{E}_{\mathbf{y}}, \mathcal{E}_{\mathbf{x}})\mathbb{P}(\mathcal{E}_{\mathbf{x}})d\mathcal{E}_{\mathbf{x}} = g(\mathcal{E}_{\mathbf{y}})$, i.e., $\mathbf{y}$ is distributed according to a spherically symmetric distribution. Let us define $\mathbf{u}_{\mathbf{y}} = \mathbf{y}/\sqrt{\mathcal{E}_{\mathbf{y}}}$. Clearly, $\mathbf{u}_{\mathbf{y}}$ is uniformly distributed on the unit m-sphere and independent of $\mathcal{E}_{\mathbf{x}}$ and $\mathcal{E}_{\mathbf{y}}$. Then, the proposition is proved by the following chain of equalities

$$\begin{aligned} I(\mathcal{E}_{\mathbf{x}}; \mathbf{y}) =& I(\mathcal{E}_{\mathbf{x}}; \mathcal{E}_{\mathbf{y}}, \mathbf{u}_{\mathbf{y}}) \\ =& I(\mathcal{E}_{\mathbf{x}}; \mathcal{E}_{\mathbf{y}}) + I(\mathcal{E}_{\mathbf{x}}; \mathbf{u}_{\mathbf{y}}|\mathcal{E}_{\mathbf{y}}) \\ =& I(\mathcal{E}_{\mathbf{x}}; \mathcal{E}_{\mathbf{y}}). \end{aligned} \tag{3.5}$$

□

By using the above results, if we transmit the spherical angle of the measurements $\mathbf{u}_{\mathbf{y}}$ instead of the measurements themselves, we can show that it is possible to increase the secrecy of the G-OTS cryptosystem, namely from spherical to *perfect secrecy*. As a matter of fact, for G-OTS measurements the energy and the spherical angle of $\mathbf{y}$ can be seen as two separate channels decoupling the information about $\mathbf{x}$. The energy of $\mathbf{y}$ carries only information about the energy of $\mathbf{x}$, whereas the spherical angle of $\mathbf{y}$ does not carry any information about $\mathbf{x}$. This normalization strategy, denoted as SG-OTS, aims at securing (in perfect secrecy sense) the measurements, and can be described as follows

$$\mathbf{u}_{\mathbf{y}} = \begin{cases} \mathbf{y}/\sqrt{\mathcal{E}_{\mathbf{y}}} & \mathcal{E}_{\mathbf{y}} > 0 \\ \mathbf{u} & \mathcal{E}_{\mathbf{y}} = 0 \end{cases} \tag{3.6}$$

where $\mathbf{u}$ is a random vector uniformly distributed on a unit radius m-sphere. This concept is formalized by the following Lemma:

Lemma 3.1 *SG-OTS measurements are perfectly secure, i.e., $\mathbb{P}(\mathbf{u}_{\mathbf{y}}|\mathbf{x}) = \mathbb{P}(\mathbf{u}_{\mathbf{y}})$.*

Proof This result easily follows from the fact that, as shown in the proof of Proposition 3.4, $\mathbf{u}_{\mathbf{y}}$ is uniformly distributed on the unit radius m-sphere irrespective of $\mathbf{x}$. □

In this section we showed that, because of the linearity of the acquisition process, the measurements leak information about the original signal. For generic sensing matrices, we have that the spherical angle of the measurements leaks only the spherical angle of the sensed signal. In the case of Gaussian sensing matrices, if no further normalization is applied, the leaked information corresponds to the energy of the original signal. Conversely, if we transmit only the spherical angle of measurements acquired using a Gaussian sensing matrix, we can achieve perfect secrecy. Interestingly, all the above results hold irrespective of the distribution and sparsity of the signal $\mathbf{x}$, which means that the above results also apply to generic embeddings using random matrices.

It is important to highlight that the energy of the original signal is the *minimum* amount of leakage we can expect from the measurements. This means that when using Gaussian i.i.d. sensing matrices it is possible to decrease the amount of leakage down to its minimum. In a nutshell, the sensing matrix has to satisfy the RIP property in order to provide the required guarantees during the recovery process. The drawback is that the energy of the original signal is guaranteed to be preserved and thus leaked. This means that for all the classes of sensing matrices which satisfy the RIP the energy will be leaked. Interestingly, as we have shown with the above results, when the sensing matrix is made of i.i.d. Gaussian entries *only* the energy is leaked. Indeed, as we will discuss in Sect. 3.3, other classes of sensing matrices do leak more information about the original signal rather than only its energy.

The above results allowed us to characterize *what* is leaked through the measurements. However, one may want to *quantify* the leakage. In other words, one may be interested to understand the precision at which an attacker can estimate the leaked information. In more detail, this latter task boils down to (1) the derivation of the attainable mean squared error (MSE) when trying to estimate the leaked energy in the case of Gaussian sensing matrices (2) characterizing the distinguishability of equal-energy signals having different spherical angles. The following sections will address this important problem.

3.2 Gaussian Sensing Matrices and Asymptotic Behavior

3.2.1 Model Definition and Security Metrics

In the previous section we analyzed the statistical properties of measurements acquired using Gaussian and arbitrary sensing matrices. Those results give insights on information that leaks through the measurements. However, they do not directly provide an indication of the kind of advantage than an attacker can obtain by observing the measurements. In the following, we will analyze such an advantage in the case of Gaussian sensing matrices. Since Gaussian measurements only provide information regarding the energy of the sensed signal, we will assume that the objective of the attacker is to estimate $\mathcal{E}_{\mathbf{x}}$. More in detail, throughout this section we will consider the following model:

- the measurements are acquired according to the OTS model;
- the sensing matrix has i.i.d. zero-mean Gaussian entries;
- the attacker has only access to the measurements $\mathbf{y}$.

In the light of the above, it is necessary to consider a new confidentiality metric which can capture the desired properties, namely it should be directly linked to the ability of the attacker of obtaining an estimated energy $\hat{\mathcal{E}}_{\mathbf{x}}$ which is close to the true energy $\mathcal{E}_{\mathbf{x}}$, when only the measurements $\mathbf{y}$ are observed.

This concept is formalized by means of the η-mean square error (MSE) metric introduced in [4].

Definition 3.3 (η*-MSE*) The measurements are said to be η-MSE secure with respect to $\mathcal{E}_{\mathbf{x}}$, if for every possible estimator $\hat{\mathcal{E}}_{\mathbf{x}}(\mathbf{y})$ of $\mathcal{E}_{\mathbf{x}}$ we have that

$$\eta_{\hat{\mathcal{E}}_{\mathbf{x}}} \triangleq \frac{\mathbb{E}\left[||\mathcal{E}_{\mathbf{x}} - \hat{\mathcal{E}}_{\mathbf{x}}(\mathbf{y})||_2^2\right]}{\sigma^2_{\mathcal{E}_{\mathbf{x}}}} \geq \eta$$

where $\mathcal{E}_{\mathbf{x}} = ||\mathbf{x}||_2^2$ and $\sigma^2_{\mathcal{E}_{\mathbf{x}}}$ is the variance of $\mathcal{E}_{\mathbf{x}}$.

It is interesting to notice that a Bayesian estimator is always at least 1-MSE secure. In fact, if no posterior information is provided, we have that the minimum MSE (MMSE) estimator of the energy of $\mathbf{x}$ is given by $\hat{\mathcal{E}}_{\mathbf{x}}(\mathbf{y}) = \mathbb{E}[\mathcal{E}_{\mathbf{x}}]$ which in turn leads to $\mathbb{E}\left[||\mathcal{E}_{\mathbf{x}} - \hat{\mathcal{E}}_{\mathbf{x}}(\mathbf{y})||_2^2\right] = \sigma^2_{\mathcal{E}_{\mathbf{x}}}$ and thus $\eta = 1$.

At this point, by recalling a result from rate-distortion theory ([10, Th. 8.6.6.]) it is possible to link the MSE of an estimator $\hat{\mathcal{E}}_{\mathbf{x}}(\mathbf{y})$ of $\mathcal{E}_{\mathbf{x}}$ relying on the measurements $\mathbf{y}$ to the mutual information between $\mathbf{y}$ and $\mathcal{E}_{\mathbf{x}}$. In more detail, it is possible to obtain the following lower bound:

$$E[||\mathcal{E}_{\mathbf{x}} - \hat{\mathcal{E}}_{\mathbf{x}}(\mathbf{y})||_2^2] \geq \frac{1}{2\pi}e^{2h(\mathcal{E}_{\mathbf{x}}|\mathbf{y})-1} = \frac{1}{2\pi}e^{2h(\mathcal{E}_{\mathbf{x}})-2I(\mathcal{E}_{\mathbf{x}};\mathbf{y})-1}. \tag{3.7}$$

From this bound it is immediate to obtain the following Lemma:

Lemma 3.2 *Measurements acquired with a generic sensing matrix are at least* η*-MSE secure with respect to* $\mathcal{E}_{\mathbf{x}}$*, where*

$$\eta = \frac{e^{2h(\mathcal{E}_{\mathbf{x}}|\mathbf{y})-1}}{2\pi\sigma^2_{\mathcal{E}_{\mathbf{x}}}}. \tag{3.8}$$

From the above results, it can be seen that if we want to obtain more meaningful results in terms of η-MSE, it is necessary to specify the distribution of both $\mathbf{x}$ and $\mathbf{\Phi}$ as this is required to evaluate the terms $h(\mathcal{E}_{\mathbf{x}}|\mathbf{y})$ and $\sigma^2_{\mathcal{E}_{\mathbf{x}}}$. While characterizing generic distributions is a tough task, we can start by fixing the distribution for the entries of the sensing matrix $\mathbf{\Phi}$. The following lemma shows an upper bounds on $I(\mathcal{E}_{\mathbf{x}}; \mathcal{E}_{\mathbf{y}})$ when the entries of $\mathbf{\Phi}$ follow a Gaussian distribution.

Lemma 3.3 *For G-OTS measurements* $I(\mathcal{E}_{\mathbf{x}}; \mathcal{E}_{\mathbf{y}})$ *can be upper bounded as*

$$I(\mathcal{E}_{\mathbf{x}}; \mathcal{E}_{\mathbf{y}}) \leq \xi(\kappa^*) - \xi\left(\frac{m}{2}\right) - \psi(\kappa^*) + \psi\left(\frac{m}{2}\right) \tag{3.9}$$

where $\xi(\kappa) \triangleq \kappa + \log(\Gamma(\kappa)) + (1-\kappa)\psi(\kappa)$, $\Gamma(\kappa) = \int_0^\infty t^{\kappa-1}e^{-t}dt$ *is the Gamma function,* $\psi(\kappa) = \frac{d\log(\Gamma(\kappa))}{d\kappa}$ *is the digamma function, and* κ^* *is the solution to the nonlinear equation* $\log(\kappa^*) - \psi(\kappa^*) = \log(m/2) - \psi(m/2) + \log(E[\mathcal{E}_{\mathbf{x}}]) - E[\log(\mathcal{E}_{\mathbf{x}})]$.

Proof See Lemma 3 in [4]. □

It is interesting to notice that the above bound does not depend on the variance of the entries $\mathbf{\Phi}_{i,j}$, however the dependence on the number of measurements m is clear.

Furthermore, by specifying a distribution for the entries of $\mathbf{x}$, it is possible to obtain closed form solutions for the value of η. Let us model $\mathbf{x}$ as an exactly k-sparse signal whose non-zero entries follow a Gaussian distribution with zero mean and variance $\sigma_{\mathbf{x}}^2$, namely $\mathbf{x}_i \sim \mathcal{N}(0, \sigma_{\mathbf{x}}^2)$. Relying on this assumption, it is possible to evaluate the term $\log(E[\mathcal{E}_{\mathbf{x}}]) - E[\log(\mathcal{E}_{\mathbf{x}})]$ in Lemma 3.3 and obtain the following result.

Corollary 3.1 *If* $\mathbf{x}$ *is an exactly k-sparse signal with i.i.d. Gaussian nonzero entries, G-OTS measurements are at least* η*-MSE secure with respect to* $\mathcal{E}_{\mathbf{x}}$*, where*

$$\eta = \frac{e^{2\xi(\frac{k}{2})+2\xi'(\frac{m}{2})-2\xi'(\kappa^*)-1}}{k\pi} \tag{3.10}$$

$\xi'(\kappa) = \xi(\kappa) - \psi(\kappa)$ *and* κ^* *satisfies* $\log(\kappa^*) - \psi(\kappa^*) = \log(m/2) - \psi(m/2) + \log(k/2) - \psi(k/2)$.

Proof See Corollary 2 in [4]. □

3.2.2 Energy Obfuscation

The definition of η-MSE security allowed us to characterize more in detail how much information can be obtained by an attacker trying to obtain estimates of $\mathcal{E}_{\mathbf{x}}$ given only the observed measurements $\mathbf{y}$. If the sensing matrix contains Gaussian i.i.d. entries, then only the energy of $\mathcal{E}_{\mathbf{x}}$ will be leaked. More specifically, it will be leaked through $\mathcal{E}_{\mathbf{y}}$ and the amount of information which it carries about $\mathcal{E}_{\mathbf{x}}$ is quantified by η.

Nonetheless, there are some cases in which revealing $\mathcal{E}_{\mathbf{x}}$ or at least, allowing an attacker to estimate it with reasonable precision, is not desirable. As an example, if we consider that the acquired signals can belong to different energy classes, then their energy is a sufficient statistic to classify those signals by only observing $\mathbf{y}$. In Sect. 3.1 we showed that by normalizing the measurements according to (3.6) it is possible to secure the measurements and prevent the information leakage. However, in order to perform a correct recovery, besides transmitting the measurements it is also necessary to transmit the energy at decryption side. This in turn would imply an encryption mechanism to protect the energy.

A viable way to avoid to transmit this information, as described in [26], is to obfuscate the energy through a random multiplication of the measurements. Since the random scalar which multiplies the measurements can be generated from the same random number generator employed for the sensing matrix entries, then there is no need to transmit additional information at decryption side. In this case the encryption model would become

$$\mathbf{z} = a\mathbf{y} = a\mathbf{\Phi}\mathbf{x} \tag{3.11}$$

where $\mathbf{z}$ are the obfuscated measurements, $a \sim \log\mathcal{N}(0, \sigma_a^2)$ is the obfuscation scalar following a Log-Normal distribution and $\mathbf{\Phi}$ is made of i.i.d. Gaussian entries. This would allow us to randomly affect only the leaked information, namely the energy, rather than changing also the spherical angle as it may happens with different transformations.

In this case, similarly to the case of unobfuscated measurements, we have that $I(\mathbf{z}; \mathbf{x}) = I(\mathcal{E}_{\mathbf{z}}, \mathcal{E}_{\mathbf{x}})$ (Lemma III.3 in [26]). More specifically the following Lemma provides an upper bound on the mutual information between the energies of the measurements and the original signal.

Lemma 3.4 *Let us assume* $\mathbb{P}(\mathcal{E}_{\mathbf{x}} = 0) = 0$ *and* $\mathbb{P}(a = 0) = 0$, *then the leakage of information of* $\mathbf{x}$ *through* $\mathbf{z}$ *is bounded by*

$$I(\mathbf{z}; \mathbf{x}) \leq \frac{1}{2} \log\left(1 + \frac{\psi_1\left(\frac{m}{2}\right) + \mathrm{var}(\log \mathcal{E}_{\mathbf{x}})}{4\sigma_a^2}\right)$$

where $\psi_1(z) = \frac{d^2}{dz^2} \ln \Gamma(z)$ *is the trigamma function.*

Proof See Lemma III.3 in [26] □

From the above result it is interesting to notice the effect of σ_a^2 on the mutual information between the energies. As expected, an increased value of σ_a^2 leads the upper bound, and thus $I(\mathbf{z}; \mathbf{x})$, to approach zero. At this point it is possible to obtain the value of the secrecy metric η which is provided by the following Lemma.

Lemma 3.5 *Obfuscated measurements are at least* η*-MSE secret with respect to* $\mathcal{E}_{\mathbf{x}}$, *where*

$$\eta = \frac{e^{h(\mathcal{E}_{\mathbf{x}}|\mathbf{z})-1}}{2\pi\sigma_{\mathcal{E}_{\mathbf{x}}}^2}.$$

Proof As for Lemma 2 in [4], by employing Theorem 8.6.6 in [10] we have that $\mathbb{E}[\|\mathcal{E}_{\mathbf{x}} - \hat{\mathcal{E}}_{\mathbf{x}}(\mathbf{z})\|_2^2] \geq \frac{1}{2\pi}e^{2h(\mathcal{E}_{\mathbf{x}})-2I(\mathcal{E}_{\mathbf{x}};\mathbf{z})-1} = \frac{1}{2\pi}e^{2h(\mathcal{E}_{\mathbf{x}}|\mathbf{z})-1}$, the result then follows from the definition of η-MSE secrecy. □

The above Lemma does not require a specific distribution for the entries of $\mathbf{x}$. Nevertheless, in order to achieve more specific results it is necessary to specify the distribution of $\mathbf{x}$ in order to evaluate the terms $h(\mathcal{E}_{\mathbf{x}}|\mathbf{z})$ and $\sigma_{\mathcal{E}_{\mathbf{x}}}^2$. As done previously, if we assume that $\mathbf{x}$ is exactly k-sparse and that the non-zero entries follow $\mathcal{N}(0, \sigma_{\mathbf{x}}^2)$, then we have the following result.

Corollary 3.2 *The minimum MSE obtainable by any estimator seeking an estimate of $\mathcal{E}_{\mathbf{x}}$ from the encrypted and obfuscated measurements $\mathbf{z}$ is given by*

$$\eta = \frac{e^{2\xi\left(\frac{k}{2}\right)-\log\left(1+\frac{\psi_1\left(\frac{m}{2}\right)+\psi_1\left(\frac{k}{2}\right)}{4\sigma_a^2}\right)-1}}{\pi k}$$

where $\xi(z) = z + \ln\Gamma(z) + (1-z)\psi(z)$ and $\psi(z)$ is the digamma function.

Proof See Corollary III.4.1 in [26]. □

From the above result it becomes then evident that a random scaling of the measurements can obfuscate the energy which is leaked when a CS cryptosystem employs Gaussian i.i.d. sensing matrices. Moreover, the variance of the obfuscation scalar affects the confidentiality which we expressed in terms of η-MSE.

We can now give the reader a feeling on the behavior and values of the above results through numerical experiments. We start by showing the behavior of the mutual information between the measurements and original signal as a function of the number of acquired measurements m. In more detail, in Fig. 3.1, we show $I(\mathcal{E}_{\mathbf{y}}; \mathcal{E}_{\mathbf{x}})$ from Lemma 3.3 and $I(\mathcal{E}_{\mathbf{z}}; \mathcal{E}_{\mathbf{x}})$ from Lemma 3.4, in the unobfuscated and obfuscated settings respectively. Since for this experiment no distribution for $\mathcal{E}_{\mathbf{x}}$ is specified and it is known that $c_0 = \log E[\mathcal{E}_{\mathbf{x}}] - E[\log \mathcal{E}_{\mathbf{x}}] \geq 0$, we fix its value to a positive constant $c_0 = 0.1$. For the same reason we fix $c_1 = \mathrm{var}(\ln \mathcal{E}_{\mathbf{x}}) = 0.2$.

It can be immediately noticed that, in the unobfuscated setting, as expected the mutual information increases with m. Intuitively, as the measurements increase, more information about $\mathcal{E}_{\mathbf{x}}$ is exposed through the energy of the measurements $\mathcal{E}_{\mathbf{y}}$. Conversely, when considering the mutual information between the obfuscated measurements and the original signal, the behavior is opposite. Even though counterintuitive, this behavior can be explained by the fact that the bound in Lemma 3.4 is loose for small m and becomes tighter as m increases. Nevertheless, it is possible to appreciate that the obfuscation is able to decrease the amount of information about $\mathcal{E}_{\mathbf{x}}$ which is leaked through the measurements. Moreover, the mutual information can be further decreased by accounting for larger values of σ_a^2.

3.2.3 Upper Bound Validation

In the next experiments, we consider a simple attack scenario in order to validate the bounds provided in Corollaries 3.1 and 3.2. An attacker having only access to $\mathbf{y}$ can estimate, as its best, the energy of the original signal $\mathcal{E}_{\mathbf{x}}$. This will allow us to compare the bounds obtained for the η-MSE metric with the result of practical estimation strategies.

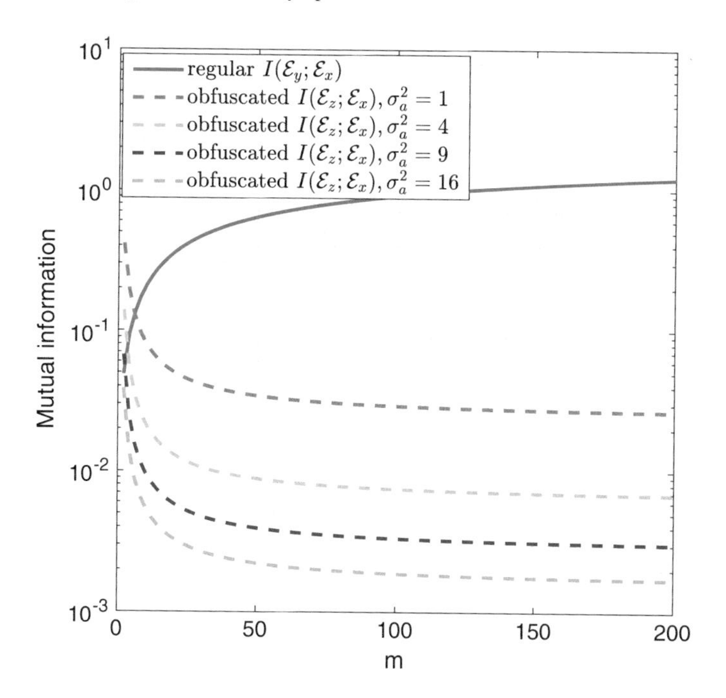

Fig. 3.1 Mutual information between the measurements and the original signal as a function of the number of measurements m. The results are shown for both obfuscated and non-obfuscated settings

Let us start by considering the estimation strategies for the case of unobfuscated measurements. By classical estimation theory results, if no prior information is available on $\mathcal{E}_{\mathbf{x}}$, we have that the variance of any unbiased estimator is always greater than the Cramér-Rao lower bound (CRLB) [19], i.e.,

$$\sigma^2_{\hat{\mathcal{E}}_{\mathbf{x}}} \geq -\frac{1}{E\left[\frac{\partial^2 L(\mathbf{y};\mathcal{E}_{\mathbf{x}})}{\partial \mathcal{E}_{\mathbf{x}}^2}\right]} \tag{3.12}$$

where $L(\mathbf{y};\mathcal{E}_{\mathbf{x}}) = \log(\mathbb{P}(\mathbf{y}|\mathcal{E}_{\mathbf{x}}))$ denotes the log-likelihood function. In the case of Gaussian sensing matrices this can be computed as

$$\sigma^2_{\hat{\mathcal{E}}_{\mathbf{x}}} \geq \frac{2\mathcal{E}_{\mathbf{x}}^2}{m}. \tag{3.13}$$

If we consider the maximum likelihood (ML) estimation strategy, we have that the estimator of $\mathcal{E}_{\mathbf{x}}$ is given by

$$\hat{\mathcal{E}}_{\mathbf{x},ML}(\mathcal{E}_{\mathbf{y}}) = \max_{\mathcal{E}_{\mathbf{x}}} \log(\mathbb{P}(\mathbf{y}|\mathcal{E}_{\mathbf{x}})) = \frac{\mathcal{E}_{\mathbf{y}}}{m\sigma^2_{\mathbf{\Phi}}}. \tag{3.14}$$

The above ML estimator is unbiased and achieves the CRLB since $E\left[\hat{\mathcal{E}}_{\mathbf{x},ML}\right] = \mathcal{E}_{\mathbf{x}}$ and $\sigma^2_{\hat{\mathcal{E}}_{\mathbf{x},ML}} = \frac{2\mathcal{E}^2_{\mathbf{x}}}{m}$. The performance of the ML estimator depends on the value of the unknown parameter $\mathcal{E}_{\mathbf{x}}$.

In order to obtain the MSE of the ML estimator under a specific distribution of $\mathcal{E}_{\mathbf{x}}$, we can observe that, for an unbiased estimator, $E_{\mathcal{E}_{\mathbf{x}},\mathbf{y}}[(\mathcal{E}_{\mathbf{x}} - \hat{\mathcal{E}}_{\mathbf{x}})^2] = E_{\mathcal{E}_{\mathbf{x}}}[\sigma^2_{\hat{\mathcal{E}}_{\mathbf{x},ML}}]$. In the case of a Gaussian k-sparse source, this results in

$$E[(\mathcal{E}_{\mathbf{x}} - \hat{\mathcal{E}}_{\mathbf{x},ML})^2] = \frac{2k(k+2)\sigma_x^4}{m} \tag{3.15}$$

from which

$$\eta_{\hat{\mathcal{E}}_{\mathbf{x},ML}(\mathcal{E}_{\mathbf{y}})} = \frac{k+2}{m}. \tag{3.16}$$

Bayesian estimators can be obtained by assuming a prior distribution for $\mathcal{E}_{\mathbf{x}}$. It is well known that in this case the MSE is minimized by the posterior mean of $\mathcal{E}_{\mathbf{x}}$, *i.e.*, $\hat{\mathcal{E}}_{\mathbf{x},MMSE} = E_{\mathcal{E}_{\mathbf{x}}|\mathbf{y}}[\mathcal{E}_{\mathbf{x}}]$ [19].

For a Gaussian k-sparse signal, it is possible to derive a closed form of the MMSE estimator as

$$\hat{\mathcal{E}}_{\mathbf{x},MMSE}(\mathcal{E}_{\mathbf{y}}) = \frac{\sigma_{\mathbf{x}}\sqrt{\mathcal{E}_{\mathbf{y}}}}{\sigma_{\mathbf{\Phi}}} \frac{K_{\frac{k}{2}-\frac{m}{2}+1}\left(\frac{\sqrt{\mathcal{E}_{\mathbf{y}}}}{\sigma_{\mathbf{\Phi}}\sigma_{\mathbf{x}}}\right)}{K_{\frac{k}{2}-\frac{m}{2}}\left(\frac{\sqrt{\mathcal{E}_{\mathbf{y}}}}{\sigma_{\mathbf{\Phi}}\sigma_{\mathbf{x}}}\right)} \tag{3.17}$$

where $K_\nu(\mathbf{x})$ denotes the modified Bessel function of the second kind of order ν. Nevertheless, a closed form solution for the MSE of this estimator cannot be obtained. In this regard it is possible to make reasonable assumptions, i.e. considering the best *linear* estimator of $\mathcal{E}_{\mathbf{y}}$ in terms of MSE that is the linear MMSE (LMMSE). The general expression of the LMMSE is $\hat{\mathcal{E}}_{\mathbf{x},LMMSE} = \mathbf{C}_{\mathcal{E}_{\mathbf{x}}\mathcal{E}_{\mathbf{y}}}\mathbf{C}^{-1}_{\mathcal{E}_{\mathbf{y}}}\left(\mathcal{E}_{\mathbf{y}} - E[\mathcal{E}_{\mathbf{y}}]\right) + E[\mathcal{E}_{\mathbf{x}}]$, where $\mathbf{C}_{\mathcal{E}_{\mathbf{y}}} = \sigma^2_{\mathcal{E}_{\mathbf{y}}}$ and $\mathbf{C}_{\mathcal{E}_{\mathbf{x}}\mathcal{E}_{\mathbf{y}}} = E[\mathcal{E}_{\mathbf{x}}\mathcal{E}_{\mathbf{y}}] - E[\mathcal{E}_{\mathbf{x}}]E[\mathcal{E}_{\mathbf{y}}]$ [19].

As previously done, we derive the LMMSE estimator for Gaussian k-sparse signals as

$$\hat{\mathcal{E}}_{\mathbf{x},LMMSE}(\mathcal{E}_{\mathbf{y}}) = \frac{\mathcal{E}_{\mathbf{y}}}{\sigma^2_{\mathbf{\Phi}}(m+k+2)} + \frac{k(k+2)\sigma_x^2}{m+k+2}. \tag{3.18}$$

Then, the MSE can be evaluated as

$$E[(\mathcal{E}_{\mathbf{x}} - \hat{\mathcal{E}}_{\mathbf{x},LMMSE})^2] = \frac{2k(k+2)\sigma_x^4}{m+k+2} \tag{3.19}$$

from which we obtain

$$\eta_{\hat{\mathcal{E}}_{\mathbf{x},LMMSE}(\mathcal{E}_{\mathbf{y}})} = \frac{k+2}{m+k+2}. \tag{3.20}$$

We now consider the case of obfuscated measurements. Since in this case ML and MMSE estimators would be more difficult to obtain in closed form, we consider as in [26] the LMMSE estimator which can be obtained as

$$\hat{\mathcal{E}}_{\mathbf{x},LMMSE}(\mathcal{E}_{\mathbf{z}}) = \frac{2\mathcal{E}_{\mathbf{z}}}{\sigma_{\mathbf{\Phi}}^2 h} + \frac{k\sigma_{\mathbf{x}}^2(h - 2me^{2\sigma_a^2})}{h} \tag{3.21}$$

where $h = 2(2 + k + m)(g + e^{2\sigma_a^2}) + mkg$ and $g = (e^{4\sigma_a^2} - 1)e^{4\sigma_a^2}$. Then, the MSE achieved by this estimator can be shown to be

$$\eta_{\hat{\mathcal{E}}_{\mathbf{x},LMMSE}(\mathcal{E}_{\mathbf{z}})} = 1 - \frac{2me^{4\sigma_a^2}}{h}. \tag{3.22}$$

In Figs. 3.2 and 3.3 we show the behavior of the lower bounds in (3.1) and (3.2) compared with the estimators we discussed above for both unobfuscated and obfuscated measurements. These values are displayed as a function of the number of measurements m for a fixed measurements sparsity ratio $\rho = m/k = 0.5$. The sparsity varies between 1 and 200 and the simulated results are averaged over 10^5 realizations for each value of k. It is important to consider that we choose a fixed measurements sparsity ratio, since it is a reasonable assumption for a correct recovery of the original

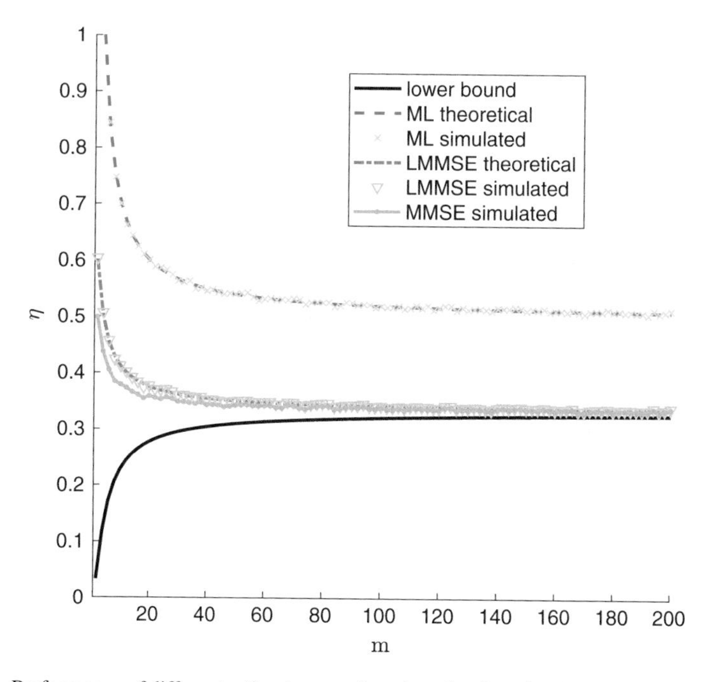

Fig. 3.2 Performance of different estimators as a function of m for a fixed measurement-to-sparsity ratio $m/k = 0.5$ when no obfuscation is considered

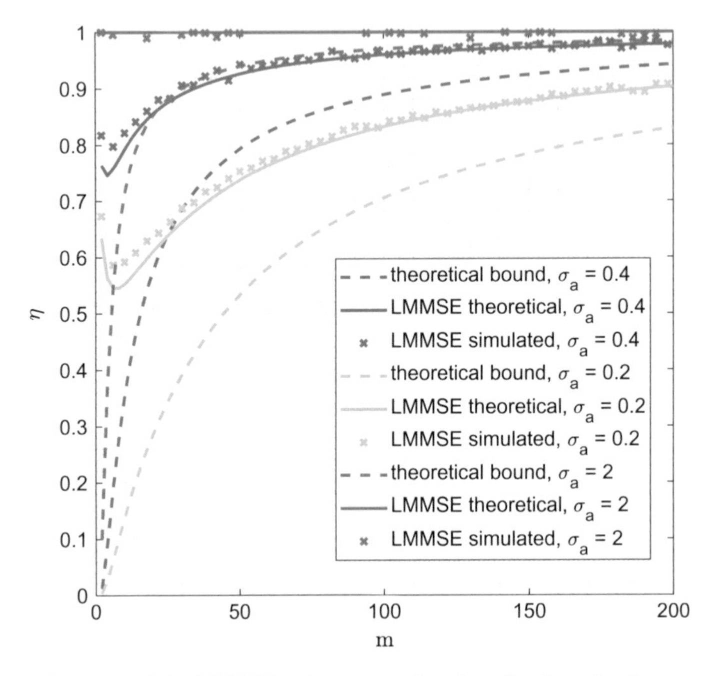

Fig. 3.3 Performance of the LMMSE estimator as a function of m for a fixed measurement-to-sparsity ratio $m/k = 0.5$ when obfuscation is applied. The results are depicted for different values of the variance of the obfuscation scaling factor σ_a

signal. It can be shown [12] that under mild conditions on the independence of the columns of $\mathbf{\Phi}$, if $m \geq 2k$ then the recovery process will converge to exact solution with high probability.

The results depicted in Figs. 3.2 and 3.3 show, as a general trend, that the ability of the considered estimator to output correct estimates increases with m until a saturation point. In fact, given a fixed sparsity k, once the minimum amount of information (measurements) required to minimize the estimation error is achieved, it will not further decrease. In addition, it is possible to appreciate that, as a general remark, the theoretical bounds and the estimation error converge as m increases. This suggests that the lower bounds on η might be loose for small values of m. Lastly, it is interesting to notice that, in case of obfuscated measurements, an increase in the obfuscator variance σ_a^2 corresponds to an increase in the estimation error.

3.2.4 Asymptotic Behavior

So far we have analyzed in non-asymptotic setting the properties of CS cryptosystems which make use of Gaussian sensing matrices. The previous results only hold for this specific distribution of the sensing matrix entries. However, because of implementation requirements, in practice one may want to use sensing matrices with distributions other than Gaussian, for which unfortunately $\mathbb{P}(\mathbf{y}|\mathbf{x}) \neq \mathbb{P}(\mathbf{y}|\mathcal{E}_{\mathbf{x}})$. Examples include antipodal distributions with entries in $\{-1, 1\}$ and circulant sensing matrices. As a matter of fact, these kinds of sensing matrices are extremely useful in practical cryptosystems because of their efficient implementation.

Even though their information leakage will be higher with respect to those with Gaussian entries, as we will discuss in detail in the following section, we may consider them in the asymptotic setting. When n, namely the size of the input signal $\mathbf{x}$ grows, by the Central Limit Theorem we have that $\mathbb{P}(\mathbf{y}|\mathbf{x})$ will tend to a multivariate Gaussian with zero mean and covariance $\sigma_{\mathbf{\Phi}}^2 \mathcal{E}_{\mathbf{x}} I_m$. This means that, as n grows, the additional leakage due to the entries of $\mathbf{\Phi}$ being non-Gaussian will decrease. In the limit of $n \to +\infty$ the only leakage will correspond to the energy of the original signal $\mathcal{E}_{\mathbf{x}}$.

This concept can be formalized using the definition of perfect secrecy in asymptotic sense introduced in [8].

Definition 3.4 (*Asymptotical spherical secrecy*) Let X be a random process whose realizations $\mathbf{x}$ have finite power $\mathcal{W}_{\mathbf{x}}$, i.e. $\lim_{n\to+\infty} \frac{1}{n} \sum_{j=0}^{n-1} x_j^2 < +\infty$ and Y be a random process whose realizations are the result of the mapping defined in (2.1). Then, a CS cryptosystem achieves asymptotical spherical secrecy if

$$\mathbb{P}(\mathbf{y}|\mathbf{x}) \underset{\mathcal{D}}{\to} \mathbb{P}(\mathbf{y}|\mathcal{W}_{\mathbf{x}})$$

holds for any realization of X and Y, where $\underset{\mathcal{D}}{\to}$ denotes the convergence in distribution as $m, n \to +\infty$.

The above definition generalizes, in asymptotic sense, the definition of spherical secrecy, namely when the measurements are secure with respect to the spherical angle of the original signal. Further, according to the above definition, when the asymptotical spherical secrecy is reached, the measurements $\mathbf{y}$ will only exhibit statistical dependence with the power of the original signal $\mathcal{W}_{\mathbf{x}}$. Indeed, the above definition allows us to characterize the information leakage of generic sensing matrices in asymptotic sense.

The following Proposition clarifies this concept from a formal point of view.

Proposition 3.5 *Let X be a random process whose realizations $\mathbf{x}$ have bounded values and have finite power $\mathcal{W}_{\mathbf{x}}$, Y be a random process whose realizations $\mathbf{y}$ are the result of the mapping in (2.1) and $\mathbf{\Phi}$ being made of i.i.d. entries with unit variance, then for $n \to +\infty$ we have*

$$\mathbb{P}\left(\frac{y_j}{\sqrt{n}}|\mathbf{x}\right) \underset{\mathcal{D}}{\to} \mathcal{N}(0, \mathcal{W}_{\mathbf{x}}).$$

Proof See Proposition 2 in [8]. □

An immediate result of the above proposition is that the results valid for Gaussian sensing matrices can be extended to generic matrices in the asymptotic sense:

Corollary 3.3 *Let X be a random process whose realizations* $\mathbf{x}$ *have bounded values and have finite power* $\mathcal{W}_{\mathbf{x}}$*, Y be a random process whose realizations* $\mathbf{y}$ *are the result of the mapping in (2.1) and* $\mathbf{\Phi}$ *being made of i.i.d. entries with finite variance, then there exists an* n^* *such that for* $n > n^*$ *the results in Lemmas 3.3 and 3.4 hold.*

Proof One may note that the results in Lemmas 3.3 and 3.4 can hold with an equal sign only if the energy of the measurements is Gaussian distributed, which is never the case. Let us consider $I(\mathcal{E}_{\mathbf{x}}; \mathcal{E}_{\mathbf{y}})$. Due to Proposition 3.5, for $n \to +\infty$ the mutual information will tend to the value obtained for Gaussian sensing matrices, which is strictly less than the value in Lemma 3.3. Hence, there should be an n^* such that for $n > n^*$ the mutual information is less than the value in Lemma 3.3. With a similar argument, the result can be proved for the case of Lemma 3.4. □

The above results highlights the fact that under the asymptotic setting it is still possible to provide some secrecy notion for generic sensing matrices. Nonetheless, one may be interested in characterizing the specific information leak, in non-asymptotic sense, of different sensing matrices. In this regard, the following section will provide a detailed analysis of such leak for a few sensing matrix classes.

3.3 Arbitrary Sensing Matrices

3.3.1 Model Definition and Security Metrics

In the previous sections we have seen that if Gaussian sensing matrices are employed, then only the energy of the original signal is leaked and thus, since the spherical angle is hidden, it is possible to achieve spherical secrecy. Furthermore, if the measurements are normalized to be unit energy, then even stronger (perfect) secrecy is achievable. All of the above results can be extended to generic sensing matrices in the asymptotic sense, i.e., when the signal dimension tends to infinity. However, for practical settings it is also important to analyze the secrecy properties of generic sensing matrices for finite values of n.

In this section we bring the reader's attention to the characterization, in the non-asymptotic sense, of the leakage due to the non-Gaussianity of the sensing matrix entries. In more detail we will consider the following model:

- the measurements are acquired according to the OTS model;
- the sensing matrix has i.i.d. entries;
- the attacker has only access to the measurements $\mathbf{y}$;
- the plaintexts have constant energy (unless differently specified).

At this point it is important to highlight that generic sensing matrices do not leak only the energy of the sensed signal, hence the secrecy definitions and the results we discussed in the previous section are not adequate to fully characterize their secrecy behavior. As a simple example, consider the case depicted in Fig. 3.4. When

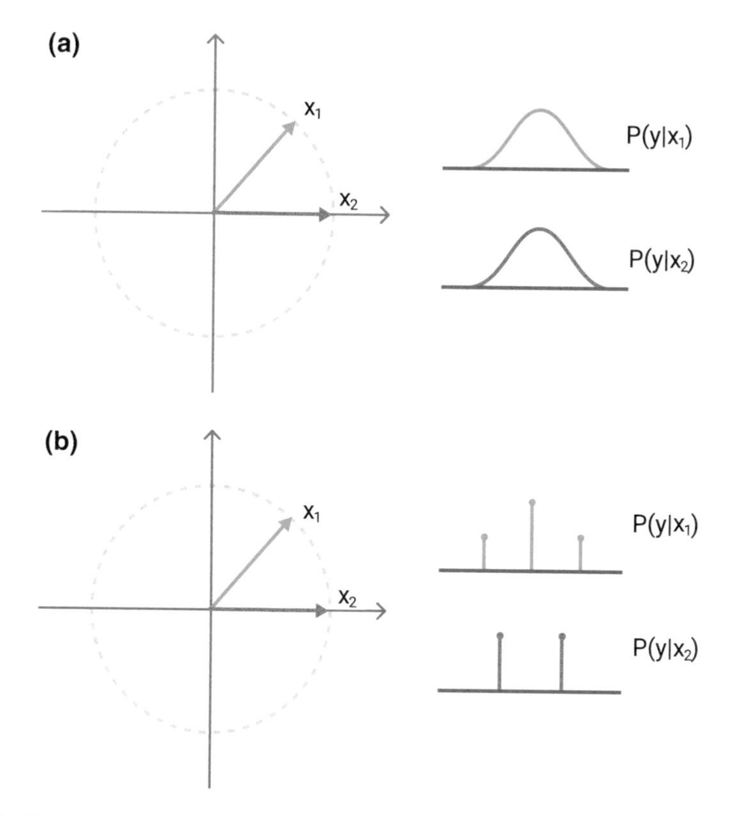

Fig. 3.4 The intuition behind spherical secrecy. When different signals are sensed using a Gaussian sensing matrix (**a**), measurements will share the same distribution. This is not case when signals are sensed using a generic distribution (**b**)

the signals $\mathbf{x}_1 = [1/\sqrt{2}, 1/\sqrt{2}]$ and $\mathbf{x}_2 = [1, 0]$ are measured projecting them onto vectors made of i.i.d Gaussian variables, measurements have the same distribution. However, if they are projected onto vectors made of i.i.d. Bernoulli variables with values in $[-1, 1]$, measurements have very different distributions.

The above problem can be addressed by employing a different metric, the ϑ-distinguishability defined in [4], which is inspired by the indistinguishability secrecy notions commonly used in cryptography [16] and recalled in Sect. 2.1. In a nutshell, this security notion characterizes the information leakage by means of a detection experiment in which, given a measurement vector $\mathbf{y}$, the goal is to detect whether the measurements come from one of two known signals $\mathbf{x}_1$ or $\mathbf{x}_2$.

More specifically, the ϑ-distinguishability experiment can be described as follows:

1. an adversary generates two signals $\mathbf{x}_1$ and $\mathbf{x}_2$ having the same length and submits them to the encoder of CS cryptosystem;
2. at encryption side, a signal $\mathbf{x}^*$ is randomly chosen from the set $\{\mathbf{x}_1, \mathbf{x}_2\}$, encrypted as $\mathbf{y} = \mathbf{\Phi}\mathbf{x}^*$ and given back to the adversary;

3. the adversary, given a detector $\mathcal{D}(\mathbf{y}) \rightarrow \{\mathbf{x}_1, \mathbf{x}_2\}$ outputs a guess on whether the encrypted signal was either $\mathbf{x}_1$ or $\mathbf{x}_2$;
4. the metrics on the success and failure of the experiment are computed as
   - probability of detection with respect to signal $\mathbf{x}_i$, $i \in \{1, 2\}$

$$P_{d,i} = \mathbb{P}(\mathcal{D}(\mathbf{y}) = \mathbf{x}_i | \mathbf{x}^* = \mathbf{x}_i)$$

   - probability of false alarm with respect to signal $\mathbf{x}_i$, $i \in \{1, 2\}$

$$P_{f,i} = \mathbb{P}(\mathcal{D}(\mathbf{y}) = \mathbf{x}_i | \mathbf{x}^* \neq \mathbf{x}_i)$$

5. given the metrics $P_{d,i}$, $P_{f,1}$, $i \in \{1, 2\}$, output of the ϑ-distinguishability experiment, it is immediate to verify $P_{d,2} = 1 - P_{f,1}$ and $P_{f,2} = 1 - P_{d,1}$, so that $P_{d,1} - P_{f,1} = P_{d,2} - P_{f,2} \triangleq P_d - P_f$.

We are now ready to state the definition of ϑ-distinguishability.

Definition 3.5 CS measurements are said to be ϑ-indistinguishable with respect to two signals $\mathbf{x}_1$ and $\mathbf{x}_2$ if for every possible detector $\mathcal{D}(\mathbf{y})$ we have

$$P_d - P_f \leq \vartheta. \tag{3.23}$$

It is interesting to note that lower values of ϑ corresponds to higher secrecy. In the limit case of $\vartheta = 0$, the above definition is equivalent to perfect secrecy, in fact having that $P_d = P_f = 0.5$ means that the detector cannot do better than random guessing. This is indeed the case where no information can be extracted out of the measurements and thus there is no information leakage.

This definition of secrecy is very general and can be applied for arbitrary sensing matrices and signal distributions. As a matter of fact, ϑ-distinguishability can be used also to characterize the ability to distinguish measurements of signals with different energies, as an alternative to η-MSE secrecy. An interested reader will find a more detailed analysis of this specific scenario in Sect. 3.3.2.

Nevertheless, a more interesting scenario is when the signals $\mathbf{x}_1$ and $\mathbf{x}_2$ do share the same energy, since in this case ϑ-distinguishability is essentially capturing the additional leakage due to the fact that the sensing matrix is not Gaussian. As shown in [4], this boils down to being able to distinguish the distributions $\mathbb{P}(\mathbf{y}|\mathbf{x}_1)$ and $\mathbb{P}(\mathbf{y}|\mathbf{x}_2)$. This means that the ϑ-distinguishability can be measured through a distance measure defined over a probability space.

In more detail, given the measurements acquired using a sensing matrix $\mathbf{\Phi}$ with arbitrarily distributed entries, it is possible to upper bound the ϑ-indistinguishability using the distributions of the measurements given the plaintext, namely $\mathbb{P}(\mathbf{y}|\mathbf{x}_1)$ and $\mathbb{P}(\mathbf{y}|\mathbf{x}_2)$. In this regard, let us recall two important probability similarity measures which will be used in the following.

Definition 3.6 (*Total Variation distance*) The Total Variation (TV) distance between two probability distributions P and Q defined over $\mathcal{X}$, is given by

$$\delta_{\mathrm{TV}}(P, Q) = \sup_{x \in \mathcal{X}} |P(x) - Q(x)|.$$

Definition 3.7 (*Kullback–Leibler divergence*) The Kullback–Leibler (KL) divergence between two probability distributions P and Q defined over $\mathcal{X}$, is given by

$$D_{\mathrm{KL}}(P||Q) = \int_{\mathcal{X}} P(x) \log \frac{P(x)}{Q(x)} dx.$$

In addition, let us recall that it is also possible to put in relation these two similarity measures through the Pinsker's inequality:

$$\delta_{TV}(P, Q) \leq \sqrt{\frac{1}{2} D_{KL}(P||Q)}. \tag{3.24}$$

As the reader has been introduced with the required similarity measures, the link existing between the ϑ-distinguishability and the TV-distance can now be presented in the following lemma.

Lemma 3.6 *OTS measurements of generic sensing matrices are at least* δ_{TV} $(\mathbf{\Phi x}_1, \mathbf{\Phi x}_2)$*-indistinguishable with respect to two signals* $\mathbf{x}_1$ *and* $\mathbf{x}_2$.

Proof The sum of error probabilities in a statistical hypothesis test can be lower bounded as [20]

$$\begin{aligned} \mathbb{P}\{\mathcal{D}(\mathbf{y}) = \mathbf{x}_2|\mathbf{x}_1\} + \mathbb{P}\{\mathcal{D}(\mathbf{y}) = \mathbf{x}_1|\mathbf{x}_2\} &= 1 - P_d + P_f \\ &\geq 1 - \delta(\mathbb{P}(\mathbf{y}|\mathbf{x}_1), \mathbb{P}(\mathbf{y}|\mathbf{x}_2)) \end{aligned} \tag{3.25}$$

from which it is possible to derive $P_d - P_f \leq \delta_{TV}(\mathbf{\Phi x}_1, \mathbf{\Phi x}_2)$. □

Let us also highlight that in the following sections most of the results will be presented in terms of upper bounds on ϑ-distinguishability or directly as distance measures between the conditional measurements distributions. As a matter of fact, being ϑ a security parameter which should ideally be as small as possible, upper bounds provide a meaningful characterization from a worst-case point of view. In fact, from security applications a worst-case analysis is mostly needed. Conversely, average case bounds are meaningful metrics which can highlight situations in which the probability of being in the worst case is extremely low and thus a higher secrecy is expected on average. Indeed, some cases in the following will require this kind of analysis.

Lastly, it is important to notice that the bounds we are going to present and which hold for generic sensing matrices are not asymptotic bounds. Nevertheless, as the reader will appreciate, we can anticipate that similarly to what we discussed in Sect. 3.2.4 as the size of signals n grows, the information leakage is reduced.

3.3.2 Generic Unstructured Sensing Matrices

Let us start the analysis by focusing on the generic case of a sensing matrix $\mathbf{\Phi}$ whose entries are i.i.d. with zero mean. As previously said, we consider plaintexts having constant energy, hence, without loss of generality, let us assume that they have unitary energy $\mathcal{E}_{\mathbf{x}_1} = \mathcal{E}_{\mathbf{x}_2} = 1$.

The first important result allows us to establish an upper bound on the TV-distance between the conditional measurements distributions, when using a generic sensing matrix $\mathbf{\Phi}$ which satisfies some assumptions as described below.

Proposition 3.6 *Let $\phi = \mathbf{\Phi}_{i,j}$ be a random variable satisfying a Poincaré inequality with constant $c > 0$, i.e. for every smooth function s with derivative s', one has $\mathrm{Var}[s(\phi)] \leq c^{-1} E[(s'(\phi))^2]$, then the TV distance between $\mathbf{\Phi x}_1$ and $\mathbf{\Phi x}_2$ is upper bounded as*

$$\begin{aligned}\delta_{TV}(\mathbf{\Phi x}_1, \mathbf{\Phi x}_2) \leq & \vartheta_{\mathbf{\Phi}}(\mathbf{x}_1, \mathbf{x}_2) \\ \triangleq & \sqrt{\frac{m D_{KL}(\phi||G)||\mathbf{x}_1||_4^4}{c + (2-c)||\mathbf{x}_1||_4^4}} + \sqrt{\frac{m D_{KL}(\phi||G)||\mathbf{x}_2||_4^4}{c + (2-c)||\mathbf{x}_2||_4^4}}\end{aligned} \tag{3.26}$$

where $D_{KL}(\phi||G) = h(G) - h(\phi)$ is the KL divergence of a Gaussian variable G with zero mean and variance $\sigma^2_{\mathbf{\Phi}}$ from ϕ.

Proof See Proposition 4 in [4]. □

The above result allows to quantify the ϑ-distinguishability when the considered signals are arbitrary and the employed sensing matrix is made of entries whose distribution satisfies the Poincaré inequality. It is interesting to consider the distributions for which this inequality holds. In [2] it was shown that it holds for log-concave probability distributions, namely those distribution in the form of $\mathbb{P}(\phi) \sim e^{-f(\phi)}$ and $f(\phi)$ convex. This means that for all Gaussian, sub-Gaussian and exponential distributions the above Proposition will hold.

Interestingly, it can also be noticed that the above proposition depends on the fourth order moments of the signals. This is indeed the rate of convergence at which the dissimilarity (in terms of KL divergence) between the distribution of the measurements (under the assumptions in Proposition 3.6) and that of a Gaussian distribution tends to zero.

Since as previously stated the above upper bound depends of the values of $||\mathbf{x}_1||_4^4$ and $||\mathbf{x}_2||_4^4$, it is interesting to analyze for which kind of signals the bound is either maximized or minimized. Recalling that we are considering signals with unitary energy, we have that the smallest achievable value for the fourth order moments is $||\mathbf{x}_1||_4^4 = ||\mathbf{x}_2||_4^4 = 1/n$. This implies that these signals have elements valued $\pm 1/\sqrt{n}$, namely $\mathbf{x}_{1,2} = [\pm 1/\sqrt{n}, \ldots, \pm 1/\sqrt{n}]$. As a result, the bound is minimized and attains its minimum given by

$$\delta_{TV}(\mathbf{\Phi x}_1, \mathbf{\Phi x}_2) \leq \sqrt{\frac{4m D_{KL}(\phi||G)}{c(n-1)+2}}.$$

In this case we have that TV distance goes to zero at least as $O(n^{-1/2})$. However, as discussed in [4], if some additional constraints on the sensing matrix are considered, it is possible to achieve a convergence rate of $O(n^{-1})$.

The other case we analyze, namely the values of $\mathbf{x}_1$ and $\mathbf{x}_2$ which maximize the bound, occur when $||\mathbf{x}_1||_4^4 = ||\mathbf{x}_2||_4^4 = 1$. This implies that both signals have only one non-zero entry which is exactly equal to 1. In this case the maximum attainable value for the bound is given by

$$\delta_{TV}(\mathbf{\Phi x}_1, \mathbf{\Phi x}_2) \leq \sqrt{2m D_{KL}(\phi||G)}.$$

At this point one may wonder how tight is the above bound with respect to the true ϑ-distinguishability. Though hard to characterize, it is immediate to notice that, if we consider two signals $\mathbf{x}_1$ and $\mathbf{x}_2$ containing the same entries (with or without permutation), the conditional distributions of the measurements will be exactly the same and thus $\vartheta = 0$. However, if $||\mathbf{x}_1||_4^4 = ||\mathbf{x}_2||_4^4 \neq 0$, the bound cannot reach zero unless ϕ follows a Gaussian distribution. The main reason behind such behavior is that the bound in (3.26) characterizes the distance of the measurements of $\mathbf{x}_1$ and $\mathbf{x}_2$ with respect to a vector of ideal measurements obtained through a Gaussian sensing matrix. Nevertheless, the bound still provides information regarding the structure of easily distinguishable signals. As a matter of fact, when $||\mathbf{x}_{1,2}||_4^4 = 1$ the signal has only a nonzero entry equal to one, which means that corresponding measurements will have the same distribution as sensing matrix entries irrespective of signal length. Conversely, $||\mathbf{x}_{1,2}||_4^4 = 1/n$ corresponds to signals having entries with equal magnitude, ensuring that the measurements will converge to a Gaussian distribution with the maximum possible speed.

Interestingly, this bound allows us to verify from a different perspective, that if the entries of the sensing matrix ϕ follow a Gaussian distribution and $\mathcal{E}_{\mathbf{x}_1} = \mathcal{E}_{\mathbf{x}_2}$, we have that $D_{KL}(\phi||G) = 0$ and thus perfect secrecy is achieved.

As a remark, let us highlight that since it is not asymptotic, the bound we presented in this section can provide meaningful results for the secrecy of a CS cryptosystem employing generic sensing matrices. In fact, if we compare it with that in Proposition 3.5 we have that the latter only claims a convergence in distribution to a Gaussian retaining only the energy (power) information of the original signal for $n \to +\infty$. This eventually leads to the spherical secrecy. In the case we are now considering, even though some assumptions are being made on sensing matrix entries ϕ, the bound can provide results in terms of ϑ-distinguishability even when the convergence of the distributions is not reached.

Furthermore, the above upper bounds can be used even in the case of normalized measurements, when for $\mathbf{x}_1$ and $\mathbf{x}_2$ we have that $\mathcal{E}_{\mathbf{x}_1} \neq \mathcal{E}_{\mathbf{x}_2} \neq 1$. Let us define the normalized measurements as $\mathbf{u}_{\mathbf{x}_i} = \mathbf{x}_i/\sqrt{\mathcal{E}_{\mathbf{x}_i}}$ and $\mathbf{u}_{\mathbf{y}_i} = \mathbf{y}_i/\sqrt{\mathcal{E}_{\mathbf{y}_i}}$, where $\mathbf{y}_i = \mathbf{\Phi x}_i$, $i = 1, 2$. Then we have the following

Corollary 3.4 *Under the hypotheses of Proposition 3.6*

$$\delta_{TV}(\mathbf{u}_{\mathbf{y}_1}, \mathbf{u}_{\mathbf{y}_2}) \leq \vartheta_{\mathbf{\Phi}}(\mathbf{u}_{\mathbf{x}_1}, \mathbf{u}_{\mathbf{x}_2}). \tag{3.27}$$

Proof See Corollary 3 in [4]. □

The above result implies that if OTS measurements are ϑ-indistinguishable with respect to equal-energy signals, then the normalized version of the same measurements is at least ϑ-indistinguishable with respect to generic signals.

Numerical Evaluation of ϑ-Distinguishability

The upper bounds of the previous section can be difficult to evaluate for some distributions commonly used in sensing matrix design. For example, for discrete distributions, e.g., Bernoulli, the term $D_{KL}(\phi||G)$ is infinite, leading to a meaningless bound.

In principle, when measurements are represented with infinite precision, sensing matrices obtained from discrete distributions will produce measurements that can only take a finite set of possible values. Since for different signals those sets will be disjunct with probability one, this will lead in general to 1-distinguishability.

In order to overcome this limitation, we can assume that, in practice, measurements will always be represented with some small approximation error. According to this model, measurements are obtained as

$$\mathbf{y} = \mathbf{\Phi}\mathbf{x} + \mathbf{n},$$

where $\mathbf{n}$ is a i.i.d. error term with zero mean and a negligible variance, independent from the other quantities. Unfortunately, in the case of generic sensing matrix distributions it is usually difficult to obtain analytical bounds for the above model.

A possible solution is to consider a numerical approximation to the upper bound. Let us consider the characteristic function of the random variable $\phi = \mathbf{\Phi}_{ij}$, defined as $f_\phi(t) = E[e^{jt\phi}]$. It is well known that the pdf of a random variable ϕ can be obtained as $\mathbb{P}(\phi) = \frac{1}{2\pi}\int_{-\infty}^{\infty} f_\phi(t)e^{-jt\phi}dt$, *i.e.*, that the characteristic function and the corresponding pdf form a Fourier transform pair. Thus, since the entries of $\mathbf{y}$ are independent, we can evaluate the characteristic function of y_i given a generic signal $\mathbf{x}$ as

$$f_{y_i|\mathbf{x}}(t) = f_{n_i}(t)\prod_{j=1}^{n} f_\phi(x_j t) \tag{3.28}$$

It is clear that, given $\mathbf{x}_1$ and $\mathbf{x}_2$, by using (3.28) it is possible to evaluate the characteristic functions $f_{y_i|\mathbf{x}_1}$ and $f_{y_i|\mathbf{x}_2}$, obtain the related probability density function through a Fourier transform, and use them to compute the KL divergence between the distributions. Finally, the term $P_d - P_f$ can be upper-bounded through the KL divergence as

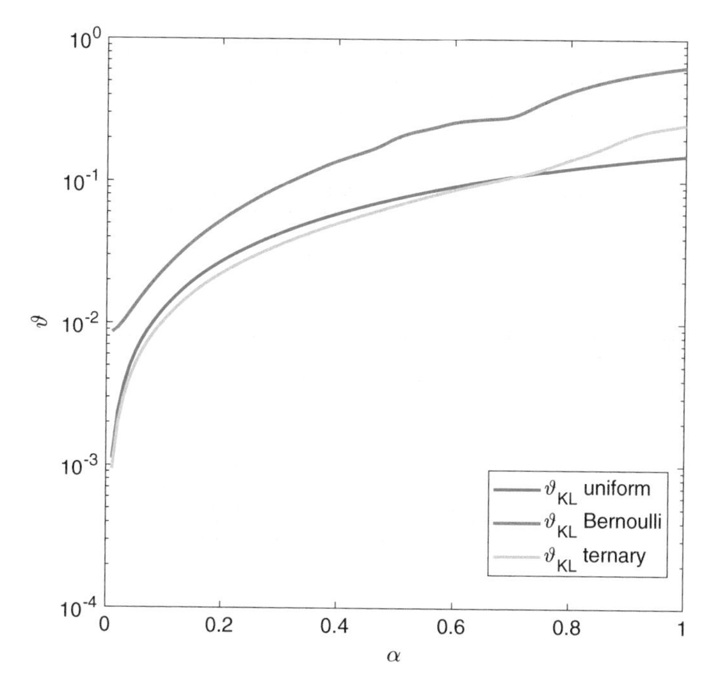

Fig. 3.5 Secrecy behavior of different sensing matrix distributions. The value of $\vartheta_{\mathbf{\Phi},\mathrm{KL}}(\mathbf{x}_1, \mathbf{x}_2)$ is plotted for $[\mathbf{x}_1]_i = 1/\sqrt{n}$ and $[\mathbf{x}_2]_i = H(\alpha)e^{-\alpha(i-1)}$ for $i = 1, \ldots, n$, where $H(\alpha)$ is a suitable value to normalize the energy of the signal

$$\begin{aligned} P_d - P_f \leq & \vartheta_{\mathbf{\Phi},\mathrm{KL}}(\mathbf{x}_1, \mathbf{x}_2) \\ \triangleq & \sqrt{\frac{m}{2} \min\left(D_{KL}([\mathbf{\Phi}\mathbf{x}_1]_i || [\mathbf{\Phi}\mathbf{x}_2]_i), D_{KL}([\mathbf{\Phi}\mathbf{x}_2]_i || [\mathbf{\Phi}\mathbf{x}_1]_i)\right)} \end{aligned} \tag{3.29}$$

where $D_{KL}([\mathbf{\Phi}\mathbf{x}_1]_i || [\mathbf{\Phi}\mathbf{x}_2]_i)$ and $D_{KL}([\mathbf{\Phi}\mathbf{x}_2]_i || [\mathbf{\Phi}\mathbf{x}_1]_i)$ are numerically evaluated.

In Fig. 3.5, we show some bounds on the ϑ-distinguishability computed for different distributions of the sensing matrix entries. Since we are considering equal energy signals, we set all the entries of $\mathbf{x}_1$ to be $1/\sqrt{n}$ and $[\mathbf{x}_2]_i = H(\alpha)e^{-\alpha(i-1)}$ for $i = 1, \ldots, n$, where $H(\alpha)$ is a suitable value to normalize the energy of the signal. It is clear that, in this case, for $\alpha = 0$ the two signals will be equal. Moreover, as discussed in the previous section, the parameter α will allow us to consider the lowest value of the bound in (3.26). We considered the entries of $\mathbf{\Phi}$ i.i.d. with zero mean and unit variance, according to the following distributions: 1) uniform; 2) Bernoulli, i.e., $\mathbb{P}\{\phi = \pm 1\} = 1/2$; 3) ternary (sparse Bernoulli) $\mathbb{P}\{\phi = \pm\sqrt{2}\} = 1/4$, $\mathbb{P}\{\phi = 0\} = 1/2$. As can be seen, Bernoulli measurements provide less secrecy compared to other distribution. Interestingly, a slightly more complex ternary distribution provides similar secrecy guarantees as a uniform distribution, as long as signal structures are not very different.

Expected Confidentiality for Constant Energy Signals

As for the results we previously discussed, the confidentiality of generic sensing matrices measured in terms of ϑ-distinguishability can be upper bounded by Eq. 3.26. Furthermore, we discussed the tightness of the bound and later showed that in some cases, e.g. when $||x_1||_4^4 = 1$ or $||x_2||_4^4 = 1$, the distinguishability can be high. Indeed, among all the possible signals with equal energy, this is the case in which worst results are reached.

A careful reader may wonder how probable such bad signal combinations are. As a matter of fact, if it is possible to show that this event is highly unlikely, then we can expect on average much lower ϑ-distinguishability and higher secrecy.

Here, we consider this problem when the signals $\mathbf{x}_1$ and $\mathbf{x}_2$ have constant energy and are uniformly distributed on an n-dimensional unit hypersphere. Before guiding the reader through the results, let us briefly discuss on the meaning of being in the worst case for the above bound. In fact, as we have seen from the bound in Eq. 3.26, there is an explicit relationship between the ϑ-distinguishability and the fourth order moments of $\mathbf{x}$. Thus, characterizing the probability of having a large value of $||\mathbf{x}||_4^4$ is essentially equivalent to characterizing the probability of being in the bad case, i.e. having a couple of signals for which $||\mathbf{x}_{1,2}||_4^4 = 1$ and the upper bound is maximized. The following lemma, from [4], addresses this problem.

Lemma 3.7 *If* $\mathbf{x}$ *is uniformly distributed on a unit n-sphere, then for* $\epsilon_1 > 0$ *and* $0 < \epsilon_2 < 1$

$$\mathbb{P}\left\{||\mathbf{x}||_4^4 \geq \frac{3+\epsilon_1}{(1-\epsilon_2)n}\right\} \leq \frac{96}{n\epsilon_1^2} + e^{-\frac{n\epsilon_2^2}{16}} \tag{3.30}$$

$$\mathbb{P}\left\{||\mathbf{x}||_4^4 \geq \frac{\epsilon_1}{(1-\epsilon_2)n}\right\} \leq \frac{5}{2}e^{-\frac{\sqrt{\epsilon_1}}{4}} + e^{-\frac{n\epsilon_2^2}{16}}. \tag{3.31}$$

Proof See Lemma 5 in [4]. □

From the above lemma it becomes clear that the probability of having large fourth order moments for the signals to be encrypted depends on n. As n becomes larger, the probability decreases.

We can now visualize the big picture of the problem we are discussing and consider the probability of having a large TV distance between the conditional measurements distributions in the case of generic sensing matrices. The Proposition below from [4], provides a solution.

Proposition 3.7 *If* $\mathbf{x}_1, \mathbf{x}_2$ *are uniformly distributed on a unit n-sphere, then with probability exceeding* $1 - \frac{192}{n\epsilon_1^2} - 2e^{-\frac{n\epsilon_2^2}{16}}$ *we have*

$$\delta_{TV}(\mathbf{\Phi}\mathbf{x}_1, \mathbf{\Phi}\mathbf{x}_2) \leq \sqrt{\frac{4m(3+\epsilon_1)D_{KL}(\phi||G)}{c(1-\epsilon_2)n + (2-c)(3+\epsilon_1)}}. \tag{3.32}$$

Moreover, with probability exceeding $1 - 5e^{-\frac{\sqrt[4]{n}}{4}} - 2e^{-\frac{n\epsilon^2}{16}}$ *we have*

$$\delta_{TV}(\mathbf{\Phi x}_1, \mathbf{\Phi x}_2) \leq \sqrt{\frac{4m D_{KL}(\phi||G)}{c(1-\epsilon)\sqrt{n}+2-c}}. \tag{3.33}$$

Proof By using the union bound, from (3.30) we have that

$$\mathbb{P}\left\{\max(||\mathbf{x}_1||_4^4, ||\mathbf{x}_2||_4^4) \leq \frac{3+\epsilon_1}{(1-\epsilon_2)n}\right\} \geq 1 - \frac{192}{n\epsilon_1^2} - 2e^{-\frac{n\epsilon_2^2}{16}}$$

which together with (3.26) immediately proves (3.32). Moreover, by setting $\epsilon_1 = \sqrt{n}$ in (3.31) we also have

$$\mathbb{P}\left\{\max(||\mathbf{x}_1||_4^4, ||\mathbf{x}_2||_4^4) \leq \frac{1}{(1-\epsilon_2)\sqrt{n}}\right\} \geq 1 - 5e^{-\frac{\sqrt[4]{n}}{4}} - 2e^{-\frac{n\epsilon_2^2}{16}}$$

which together with (3.26) immediately proves (3.33). □

Interestingly, the above result states that as n grows it becomes less likely to have large TV distance between the conditional distribution of the measurements. From another perspective, the probabilities $\mathbb{P}(\mathbf{y}|\mathbf{x}_1)$ and $\mathbb{P}(\mathbf{y}|\mathbf{x}_2)$ are likely to be close when n is large.

At this point, since the TV-distance is directly related to the ϑ-distinguishability, and by summarizing the above results we have the following Corollary.

Corollary 3.5 *If the hypotheses of the above theorems hold, then with probability exceeding* $1 - O(n^{-1})$ *we have that OTS measurements are at least* $O\left(\sqrt{\frac{m}{n}}\right)$*-indistinguishable, whereas with probability exceeding* $1 - O(e^{-\sqrt[4]{n}})$ *we have that OTS measurements are at least* $O\left(\sqrt{\frac{m}{\sqrt{n}}}\right)$*-indistinguishable.*

As a last remark, before moving to more specific cases, we can state that generic sensing matrices which follow some assumptions, namely i.i.d. entries following exponential or sub-Gaussian distributions, can have very different secrecy behaviors depending on the properties of acquired signals. Also, even though these signals can have a sufficiently large dimension that lowers the probability of information leakage, this setting will only provide a weak notion of secrecy with respect to conventional security definitions. For example, if an adversary was allowed to try a polynomial number of guesses on measurements of the same signal, he/she would have significant probability of detecting the correct signal. Nevertheless, in many practical CS settings such a scenario is quite unlikely, so the results in Corollary 3.5 can still provide adequate secrecy.

Expected Confidentiality for k-sparse Signals

As we considered for the case of the Gaussian sensing matrices in Sect. 3.2.1, we discuss the particular setting in which the signal to be acquired is exactly k-sparse in some domain. In more detail, as done in [4], let us consider the class of signals expressed as $\mathbf{x} = \mathbf{\Psi}\boldsymbol{\theta}$, where $\boldsymbol{\theta}$ has k nonzero components uniformly distributed on the unit k-sphere and $\mathbf{\Psi}$ is an orthonormal basis. Also, let us remind that this is a standard setting in CS acquisition, where the signal to be acquired is oftentimes not directly sparse in the acquisition domain, but rather it is sparse in some other domain. In this case, in the space spanned by the orthonormal basis $\mathbf{\Psi}$.

It is immediate to notice that such signals lie on the unitary n-dimensional hypersphere. In the specific case of $\mathbf{\Psi}$ being an identity matrix $\mathbf{I}_n$ the problem can be cast into an equivalent problem in which the signal is dense and uniformly distributed on the unitary k-dimensional hypersphere. As a result of the bounds discussed in Sect. 3.3.2, we have that $\delta_{TV}(\mathbf{\Phi x}_1, \mathbf{\Phi x}_2) = O(\sqrt{\frac{m}{k}})$. If the measurements should guarantee signal recovery, it is immediate to verify $O(\sqrt{\frac{m}{k}}) = O(1)$, showing that it is not possible to have a vanishing TV distance between the conditional distribution of measurements of different signals. Intuitively, this can be explained by the fact that k-sparse vectors on the unit n-sphere tend to have a larger fourth-order moment than dense vectors. As highlighted in [31], a similar result is obtained when using Haar matrices as sparsity basis.

Conversely, it is possible to envision that, if the basis $\mathbf{\Psi}$ has a different structure, then the bounds which hold for signals uniformly distributed on the n-dimensional hypersphere will still hold. In fact, in [4] the authors show that if some conditions on the moments of the entries of the sparsity basis are satisfied, then the TV distance between the conditional distribution of measurements of different signals, and thus the ϑ-distinguishability, can vanish when n increases. More specifically, if the columns ψ_i of $\mathbf{\Psi}$ satisfy $||\psi_i||_4^4 \leq \frac{C}{n}$ where C is some constant independent of n, by using Cauchy-Schwarz inequality it is easy to prove that $||\mathbf{x}||_4^4 \leq \frac{Ck^2}{n}$. Hence, if $\frac{k^2}{n} \to 0$ we have that the TV distance vanish for this class of orthonormal basis. Let us also note that examples of suitable basis which satisfy these requirements are the DCT and Walsh-Hadamard (WH) bases.

We guided the reader through the characterization of the secrecy properties of CS cryptosystems making use of generic sensing matrices. As we have seen, the results discussed so far show that the secrecy notion provided by a CS cryptosystem not relying on Gaussian sensing matrices can be very weak in the worst case, even if is usually acceptable with very high probability in the average case. We further considered the specific case of a k-sparse signal, showing that in this case the achieved secrecy depends on the sparsity basis. Popular basis, like DCT and WH, provide acceptable secrecy guarantees. Unfortunately, for some specific choices of orthonormal basis we cannot have any secrecy guarantee regardless of the size of the signals. This is to warn the reader that, if non-Gaussian sensing matrices are considered, then it is necessary to proceed with a careful analysis which also takes into account the sparsity basis employed and the prior on the signal to be acquired.

Confidentiality for Unequal Energy Signals

We previously showed the distinguishability behavior of equal energy signals acquired by means of generic sensing matrices. We also highlighted that, in the case of $\mathcal{E}_{\mathbf{x}_1} \neq \mathcal{E}_{\mathbf{x}_2}$, then perfect indistiguishability cannot be achieved because the distributions of the measurements conditioned on the plaintexts will not be equal.

Nonetheless, one might be interested to gain a deeper understanding on how sensitive is the distinguishability, and thus the TV distance between the measurements distribution, to the variations in minimum energy ratio defined as $\gamma = \mathcal{E}_{\mathbf{x}_{\min}}/\mathcal{E}_{\mathbf{x}_{\max}}$. Intuitively, if the ratio is exactly 1, then indistiguishability might be reached. Conversely, if $\gamma < 1$ it becomes interesting to characterize how far the actual distinguishability is from the perfect 0-distinguishability. Moreover, it is also interesting to analyze if an additional sensing noise corrupting the measurements has any effect on the distinguishability.

The analysis, which points to the results in [31], is restricted to cases of $\mathbf{\Phi}$ containing either i.i.d. Gaussian or Bernoulli entries. We will refer to the latter setting as B-OTS, following the same naming convention as for G-OTS.

The acquisition model we are now considering is given by:

$$\mathbf{y} = \mathbf{\Phi}\mathbf{x} + \mathbf{n} \tag{3.34}$$

where $\mathbf{n} \sim \mathcal{N}(0, \sigma_{\mathbf{n}}^2)$ is the measurements noise. Along with the energy ratio we defined above, let us also define the maximum plaintext-to-noise ratio as $\mathrm{PNR}_{\max} = \mathcal{E}_{\mathbf{x}_{\max}}/m\sigma_{\mathbf{n}}^2$ which quantifies how strong is the noise power with respect to the maximum energy achievable for the considered class of signals.

Let us start the analysis by focusing on the G-OTS model. Specifically, we have the following useful theorem.

Theorem 3.1 (Theorem 1 in [31]) *The worst-case lower and upper bounds on the TV distance between $\mathbb{P}(\mathbf{y}|\mathbf{x}_1)$ and $\mathbb{P}(\mathbf{y}|\mathbf{x}_2)$ are given by*

$$\delta_{TV,\mathrm{low}}(\mathbf{\Phi}\mathbf{x}_1, \mathbf{\Phi}\mathbf{x}_2) = 1 - \left(\frac{4\gamma_e}{(\gamma_e+1)^2}\right)^{\frac{m}{4}} \tag{3.35}$$

for the lower bound and

$$\delta_{TV,\mathrm{up}}(\mathbf{\Phi}\mathbf{x}_1, \mathbf{\Phi}\mathbf{x}_2) = 1 - \sqrt{\left(\frac{4\gamma_e}{(\gamma_e+1)^2}\right)^{\frac{m}{3}}} \tag{3.36}$$

for the upper bound, and where $\gamma_e = \frac{1+\gamma\mathrm{PNR}_{\max}}{1+\mathrm{PNR}_{\max}}$

Proof See Theorem 1 in [31]. □

From the above theorem it can be seen that the distinguishability depends on both the number of the measurements m and on the value of γ_e which accounts for the maximum to minimum energy ratio and the fact that the measurements are noisy. The indistinguishability is reached for $\gamma_e = 1$ which can be obtained when the signals have equal energy and thus $\gamma = 1$. Moreover, it is also interesting to notice that, as the $\mathrm{PNR}_{\max}$ decreases, the TV distance between the conditional measurements distributions becomes less sensitive to the minimum energy ratio γ. In other words, if the additive measurements noise is strong enough with respect to the maximum signal energy, then it can act as an obfuscator and improve the distinguishability.

Nonetheless, it is important to notice that, since by assumption a CS cryptosystem have to provide decryption capabilities, and since the probability of a successful recovery decreases with the noise power, measurements which exhibit very low $\mathrm{PNR}_{\max}$ cannot be considered a viable option.

Further, it is interesting to highlight that in this case, similarly to the results on the η-MSE in Sect. 3.2.1, we have an explicit dependency on the number of measurements m. Though, being different metrics, the dependency is not the same.

As shown by Corollary 1 in [31], the bounds of Theorem 3.1 can be included into the ϑ-distinguishability definition and used to obtain that the probability of detection satisfies

$$P_d \leq \frac{1}{2} + \frac{1}{2}\sqrt{1 - \left(\frac{4\gamma_e}{(\gamma_e + 1)^2}\right)^{\frac{m}{2}}}. \tag{3.37}$$

This makes evident that perfect indistinguishability cannot be reached since the second term in the above equation cannot reach zero unless the two signals have the same energy. Moreover, since this latter term does not depend on n, it cannot vanish for large signals. Nevertheless, if some constraints on the energy ratio are considered (as in Theorem 2 in [31]) a dependency on n can be included in the second term of (3.37) which can vanish (linearly) in n.

As an example, in Fig. 3.6 we show upper and lower bounds on the TV distance for different values of the minimum energy ratio γ, for $\mathrm{PNR}_{\max} = 0$ dB and $\mathrm{PNR}_{\max} =$ 10 dB. It is evident that, even for values of the minimum energy ratio slightly less than one, the distinguishability of signals can be significant. Moreover, additive noise tend to increase the secrecy of the system, but the effect can be quite limited in the case of unequal energy signals.

Interestingly, the above results can also be extended to the case of sensing matrices following a Bernoulli distribution (B-OTS) and, in principle, to other sensing matrix distributions. Under the assumption of $\mathbf{x}_i$ being dense (and thus sparse in some suitable orthonormal basis as discussed in Sect. 3.3.2) it is possible to obtain first-order approximations of the upper and lower worst-case bounds. In this case, the dependency on n is explicit and it can be shown that, if $\gamma = 1$ and $n \to +\infty$, B-OTS can achieve indistinguishability.

3.3.3 *Circulant Sensing Matrices*

Up to this point we focused on sensing matrices with specific or generic distributions which had in common the fact that all the entries are independent and identically distributed. However, some applications may benefit or even require sensing matrices whose entries are not independent but exhibit some correlation. Throughout this section we will examine the specific case of Gaussian circulant sensing matrices.

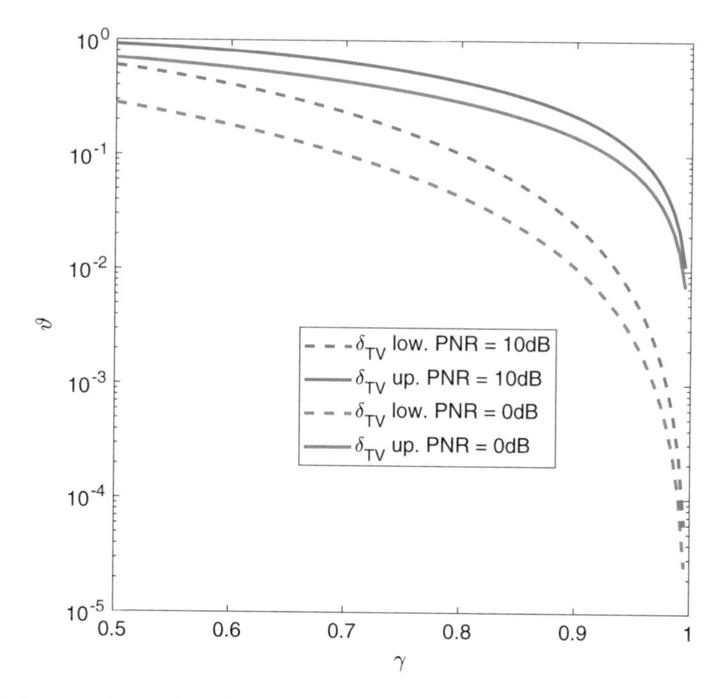

Fig. 3.6 Lower and upper bounds on the TV distance between $\mathbb{P}(\mathbf{y}|\mathbf{x}_1)$ and $\mathbb{P}(\mathbf{y}|\mathbf{x}_2)$ as a function of the minimum energy ratio γ, for $\text{PNR}_{\text{max}} = 0$ dB and $\text{PNR}_{\text{max}} = 10$ dB

The structure of a Gaussian circulant sensing matrix can be defined as follows:

$$\mathbf{\Phi} = \begin{bmatrix} \phi_1 & \phi_2 & \phi_3 & \phi_4 & \cdots & \phi_n \\ \phi_n & \phi_1 & \phi_2 & \phi_3 & \cdots & \phi_{n-1} \\ & & & \vdots & & \\ \phi_{n-m+2} & \phi_{n-m+3} & \phi_{n-m+4} & \phi_{n-m+5} & \cdots & \phi_{n-m+1} \end{bmatrix} \quad (3.38)$$

where the entries of the first row $\boldsymbol{\phi} = [\phi_1 \ \phi_2 \ \phi_3 \ \ldots \ \phi_n]$ are i.i.d. following a Gaussian distribution. Here, the first row contains all the elements which, shifted m times at each row will be used to generate the full sensing matrix.

The most evident advantage of this sensing matrix class is that, in practice, only n random entries needs to be generated in comparison to standard sensing matrices which require the generation of mn entries. Furthermore, as shown in [30] for signals which are directly sparse in the acquisition domain, it has comparable recovery performance with respect to standard sensing matrices. In the other cases the performance are worse, but still reasonable.

Another undoubtful advantage, when it comes to practical implementations, is the ability to perform fast acquisition. As a matter of fact, once the matrix has been generated, the most costly operation is related to matrix vector multiplication acquisition. More specifically, a circulant matrix $\mathbf{\Phi}$ can be diagonalized as:

$$\mathbf{\Phi} = \mathbf{P}\mathbf{F}^*\mathbf{\Lambda}\mathbf{F} \tag{3.39}$$

where $\mathbf{F}$ is the Discrete Fourier Transform (DFT) matrix and $\mathbf{P}$ is an $m \times n$ matrix that selects the first m elements of an n-dimensional vector and $\mathbf{\Lambda}$ is a diagonal matrix. This means that the acquisition process can be implemented in an efficient way by means of FFT algorithms.

Given all the aforementioned advantages, it is interesting to discuss which kind of secrecy notion can be provided by a cryptosystem employing circulant sensing matrices. At first, as done in [5], let us consider a more general definition of circulant sensing matrix:

$$\mathbf{\Phi} = \mathbf{P}\mathbf{F}^*\mathbf{\Lambda}\mathbf{F}\mathbf{R} \tag{3.40}$$

where $\mathbf{R}$ is a matrix acting as a generic scrambling operator.

Let us start the discussion considering the properties of a simple OTS-circulant cryptosystem, which we name OTS-C for short. The encryption process can be defined by $\mathbf{y} = \mathbf{\Phi}\mathbf{x}$, where $\mathbf{\Phi}$ is a circulant sensing matrix as defined in (3.40), the matrix $\mathbf{P}$ is public and $\mathbf{R}$ is an identity matrix.

As will become clearer in the following, let us also define $\mathbf{C}_\mathbf{x}$ as the circular autocorrelation matrix of $\mathbf{x}$, that is, $[\mathbf{C}_\mathbf{x}]_{ij} = \sum_{r=1}^{n} \mathbf{x}_r \mathbf{x}_{r+i-j \mod n}$, for $i, j = 1, \ldots, n$. It can be easily verified that $\mathbf{C}_\mathbf{x}$ is a Toeplitz matrix and that its diagonal elements are equal to $\mathcal{E}_\mathbf{x} = \mathbf{x}^T\mathbf{x}$. Moreover, let us consider a OTS-C cryptosystem where the entries of $\boldsymbol{\phi}$ are i.i.d. zero-mean Gaussian which we will refer to as G-OTS-C. Then, we have the following result.

Proposition 3.8 (Proposition 1 in [5]) *If ϕ_i, $i = 1, \ldots, n$, are i.i.d. zero-mean Gaussian variables, then for a G-OTS-C it holds that $\mathbb{P}(\mathbf{y}|\mathbf{x}) = \mathbb{P}(\mathbf{y}|\mathbf{P}\mathbf{C}_\mathbf{x}\mathbf{P}^T)$.*

Proof Let us consider the probability distribution of measurements $\mathbb{P}(\mathbf{y}|\mathbf{x})$ conditioned on the plaintext $\mathbf{x}$. Since ϕ_i are Gaussian, we have that $\mathbb{P}(\mathbf{y}|\mathbf{x})$ is a multivariate Gaussian distribution with mean $\mu_{\mathbf{y}|\mathbf{x}}$ and covariance matrix $\mathbf{C}_{\mathbf{y}|\mathbf{x}}$. Then, it is immediate to find $\mu_{\mathbf{y}|\mathbf{x}} = E[\mathbf{y}|\mathbf{x}] = E[A\mathbf{\Phi}]\mathbf{x} = 0$, whereas we have

$$\begin{aligned}
\mathbf{C}_{\mathbf{y}|\mathbf{x}} =& E[\mathbf{\Phi}\mathbf{x}\mathbf{x}^T\mathbf{\Phi}^T] = E[\mathbf{P}\mathbf{F}^*\mathbf{\Lambda}(\mathbf{F}\mathbf{x})(\mathbf{F}\mathbf{x})^*\mathbf{\Lambda}^H\mathbf{F}\mathbf{P}^T] \\
=& n\mathbf{P}\mathbf{F}^H \mathrm{diag}\{\mathbf{F}\mathbf{x}\} E[(\mathbf{F}^H\boldsymbol{\phi})(\mathbf{F}^H\boldsymbol{\phi})^H] \\
& \times \mathrm{diag}\{\mathbf{F}\mathbf{x}\}^H\mathbf{F}\mathbf{P}^T \\
=& n\mathbf{P}\mathbf{F}^H \mathrm{diag}\{\mathbf{F}\mathbf{x}\}\mathbf{F}^H E[\mathbf{\Phi}\mathbf{\Phi}^T]\mathbf{F}\mathrm{diag}\{\mathbf{F}\mathbf{x}\}^H\mathbf{F}\mathbf{P}^T \\
=& n\sigma_{\mathbf{\Phi}}^2 \mathbf{P}\mathbf{F}^H|\mathrm{diag}\{\mathbf{F}\mathbf{x}\}|^2\mathbf{F}\mathbf{P}^T = \sigma_{\mathbf{\Phi}}^2\mathbf{P}\mathbf{C}_\mathbf{x}\mathbf{P}^T
\end{aligned} \tag{3.41}$$

where $\mathrm{diag}\{\mathbf{v}\}$ denotes a diagonal matrix defined by vector $\mathbf{v}$. By definition, we have that $\mathbf{\Lambda} = \sqrt{n} \cdot \mathrm{diag}\{\mathbf{F}^H\boldsymbol{\phi}\}$ and that $\mathrm{diag}\{\mathbf{u}\}\mathbf{v} = \mathrm{diag}\{\mathbf{v}\}\mathbf{u}$. Moreover, we assume that the entries of $\boldsymbol{\phi}$ have variance $\sigma_{\mathbf{\Phi}}^2$. It follows that the measurements $\mathbf{y}$ depends on $\mathbf{x}$ only through the autocorrelation $\mathbf{P}\mathbf{C}_\mathbf{x}\mathbf{P}^T$, namely $\mathbb{P}(\mathbf{y}|\mathbf{x}) = \mathbb{P}(\mathbf{y}|\mathbf{P}\mathbf{C}_\mathbf{x}\mathbf{P}^T)$. □

The interesting result from the above proposition is that a G-OTS-C cryptosystem with Gaussian entries will reveal the partial circular autocorrelation matrix of the original signal $\mathbf{x}$. Moreover, it can be noted that the above result requires no assumptions on the sparsity k.

The first remark we can make is that, since by using circulant matrices we are introducing structure in the sensing process, the same structure is exposed through the measurements. As as matter of fact, while for unstructured G-OTS the probability of the measurements conditioned on $\mathbf{x}$ shows only a dependence on the energy, i.e. $\mathbb{P}(\mathbf{y}|\mathcal{E}_{\mathbf{x}})$, in the case of G-OTS-C the dependency given by $\mathbb{P}(\mathbf{y}|\mathbf{x}) = \mathbb{P}(\mathbf{y}|\mathbf{P}\mathbf{C}_{\mathbf{x}}\mathbf{P}^T)$ shows an increased information leakage. It is important to highlight that, while this behavior is undesirable for secrecy purposes, it can be exploited to perform signal processing operations in the compressed domain as discussed in [28, 29].

Let us recall that the general structure of the sensing matrix in OTS-C setting as defined in (3.40) also contains a selection matrix $\mathbf{P}$ and a random matrix $\mathbf{R}$. Interestingly, this gives us some degrees of freedom by allowing us to introduce additional randomness in the measurements and eventually decrease the statistical dependency between $\mathbf{y}$ and $\mathbf{x}$. In this regard let us consider the additional scenarios which can be envisioned:

1. **Gaussian OTS with single randomization (G-OTS-R1)**: the selection matrix $\mathbf{P}$ is kept secret and the selection of m out of n entries is randomly drawn from a uniform distribution. In this case, the randomization matrix $\mathbf{R}$ is the identity matrix and it is public.
2. **Gaussian OTS with double randomization (G-OTS-R2)**: the selection matrix $\mathbf{P}$ is defined as for G-OTS-R1. However, the randomization matrix $\mathbf{R}$ is a diagonal matrix whose non-zero elements are i.i.d. and follow a Rademacher distribution. The result is a random flipping of the signs for the elements in $\mathbf{x}$. Clearly, the matrix $\mathbf{R}$ is kept secret.

In the following we will discuss the security properties of OTS cryptosystem with i.i.d. Gaussian circulant sensing matrices without randomization (G-OTS-C) and with single (G-OTS-R1) and double randomization (G-OTS-R2).

Confidentiality of G-OTS-C

In this case no additional randomization is considered, apart from the one intrinsic in the acquisition process. As previously done for generic sensing matrices, we characterize the secrecy of such cryptosystem by means of ϑ-distinguishability.

Theorem 3.2 (Proposition 2 in [5]) *A G-OTS-C cryptosystem is at least $\vartheta_C(\mathbf{x}_1, \mathbf{x}_2)$-indistinguishable w.r.t. $\mathbf{x}_1$, $\mathbf{x}_2$, where*

$$\vartheta_C(\mathbf{x}_1, \mathbf{x}_2) = \frac{1}{2}\sqrt{\log\frac{|\mathbf{C}_2|}{|\mathbf{C}_1|} + \mathrm{Tr}(\mathbf{C}_2^{-1}\mathbf{C}_1) - m} \tag{3.42}$$

and $\mathbf{C}_h = \mathbf{P}\mathbf{C}_{\mathbf{x}_h}\mathbf{P}^T$, for $h = 1, 2$.

Proof Thanks to Proposition 3.8, we have that $\mathbb{P}(\mathbf{y}|\mathbf{x}_h) = \mathcal{N}(0, \sigma_{\mathbf{\Phi}}^2\mathbf{C}_h)$. Hence, the Kullback–Leibler (KL) divergence between $\mathbb{P}(\mathbf{y}|\mathbf{x}_1)$ and $\mathbb{P}(\mathbf{y}|\mathbf{x}_2)$ can be expressed as [11]

$$D_{KL}(\Phi\mathbf{x}_1|\Phi\mathbf{x}_2) = \frac{1}{2}\left[\log\frac{|\mathbf{C}_2|}{|\mathbf{C}_1|} + \mathrm{Tr}(\mathbf{C}_2^{-1}\mathbf{C}_1) - m\right]. \tag{3.43}$$

The result then follows from Pinsker's inequality. □

From the above theorem it can be clearly seen that the distinguishability depends on 1) the relationship between the partial autocorrelation matrices of the two signals, namely $\mathbf{C}_h,\ h = \{1, 2\}$ and 2) on the number of measurements m. Clearly, if $\mathbf{C}_{\mathbf{x}_1} = \mathbf{C}_{\mathbf{x}_2} = \mathbf{I}$, then perfect secrecy can be achieved, namely $\vartheta = 0$. In fact, this corresponds to the case of equal energy signals and uncorrelated entries, which we analyzed in Sect. 3.2.1.

It is also interesting to notice that because of the structure of the sensing matrix, additional assumptions are needed in order to achieve perfect secrecy. As a matter of fact, differently from the results in Sect. 3.2.1, here we do also require to have signals whose entries are uncorrelated. More specifically, the entries selected by the matrix $\mathbf{P}$ should not exhibit correlation.

Furthermore, from the results in [5], we have that the ϑ-distinguishability decreases for large values n when the signals are sparse in the acquisition domain. Conversely, when the sparsity is considered in the DCT domain, the distinguishability increases with n. Indeed, if the sparsity k is kept fixed, but the signal size is increased, the correlation in the acquisition domain will also increase.

Confidentiality of G-OTS-R1

We continue our discussion by introducing some additional randomness: the selection matrix $\mathbf{P}$ uniformly selects m entries at random and is kept secret. From [6] we have the following result in terms of ϑ-distinguishability.

Theorem 3.3 *A G-OTS-R1 cryptosystem is at least $\vartheta_{R1}(\mathbf{x}_1, \mathbf{x}_2)$-indistinguishable w.r.t. $\mathbf{x}_1$, $\mathbf{x}_2$, where*

$$\vartheta_{R1}(\mathbf{x}_1, \mathbf{x}_2) = \sqrt{\frac{1}{4N_P^2}\sum_{r1,r2}\left[\log\frac{|\mathbf{C}_{2,r2}|}{|\mathbf{C}_{1,r1}|} + \mathrm{Tr}(\mathbf{C}_{2,r2}^{-1}\mathbf{C}_{1,r1})\right] - \frac{m}{4}} \tag{3.44}$$

$\mathbf{C}_{h,r} = \mathbf{P}_r\mathbf{C}_{x_h}\mathbf{P}_r^T$, for $h = 1, 2$, and $N_P = n!/(n-m)!$ denotes the number of possible selection matrices.

Proof In this case we have that $\mathbb{P}(\mathbf{y}|\mathbf{x}_h)$ can be expressed as a mixture of Gaussian distributions. The KL divergence between two mixture distributions $\mathbb{P}_i = \sum_r w_{h,r}\mathbb{P}_{h,r}$, $h = 1, 2$, can be upper bounded using the following convexity bound [17]

$$D_{KL}(\mathbb{P}_1||\mathbb{P}_2) \leq \sum_{r1,r2} w_{1,r1}w_{2,r2}D_{KL}(\mathbb{P}_{1,r1}||\mathbb{P}_{2,r2}). \tag{3.45}$$

Hence, the result can be easily obtained by considering that $w_{1,r} = w_{2,r} = N_P^{-1}$ and then applying Pinsker's inequality to the upper bound on the KL divergence. □

From the above theorem, it can be seen that in this case the ϑ-distinguishability upper bound depends on all the possible random selection matrices $\mathbf{P}_r$ which need to be evaluated in its computation. This implies that for large values of m and n, the computation of the above bound can become prohibitive. As suggested in [5], a possible approach to follow is to estimate the bound by means of Monte Carlo integration.

Alternatively, following the suggestion in [17], we can approximate the KL divergence between the two mixture distributions using the KL divergence of two multivariate Gaussian distributions having the same mean and covariance matrix. In this case, the covariance matrix of the involved mixture distributions has a very peculiar form, since

$$[\mathbf{C}_h]_{ij} = \sum_{r=1}^{N_P} \frac{1}{N_P}[\mathbf{C}_{h,r}]_{ij} = \begin{cases} \sigma_{\mathbf{\Phi}}^2 \mathcal{E}_{\mathbf{x}_h} & i = j \\ \sigma_{\mathbf{\Phi}}^2 \sum_{s \neq t} \mathbf{x}_{h,s}\mathbf{x}_{h,t} & i \neq j \end{cases} \tag{3.46}$$

for $h = 1, 2$. The above covariance matrix can be expressed in a compact form as $\mathbf{C}_h = \alpha_h \mathbf{I}_m + \beta_h \mathbb{1}\mathbb{1}^T$, where we define $\alpha_h = \frac{\sigma_{\mathbf{\Phi}}^2}{n-1}(n\mathcal{E}_{x_h} - (\mathbb{1}^T\mathbf{x}_h)^2)$ and $\beta_h = \frac{\sigma_{\mathbf{\Phi}}^2}{n-1}((\mathbb{1}^T\mathbf{x}_h)^2 - \mathcal{E}_{\mathbf{x}_h})$. According to the above representation, the KL divergence between $\mathbb{P}(\mathbf{y}|\mathbf{x}_1)$ and $\mathbb{P}(\mathbf{y}|\mathbf{x}_2)$ can be approximated as

$$\begin{aligned} D_{KL}(\mathbb{P}_1||\mathbb{P}_2) \approx & \frac{1}{2}\left[\log \frac{\alpha_2^{m-1}(\alpha_2 + m\beta_2)}{\alpha_1^{m-1}(\alpha_1 + m\beta_1)} + \frac{m\alpha_2(\alpha_1+\beta_1) + m(m-1)\alpha_1\beta_2}{\alpha_2(\alpha_2 + m\beta_2)} - m\right] \\ \triangleq & \tilde{D}_{KL}(\mathbf{x}_1, \mathbf{x}_2). \end{aligned} \tag{3.47}$$

In this way it is possible to define an *approximate* security metric which is defined as

$$\vartheta'_{R1}(\mathbf{x}_1, \mathbf{x}_2) = \sqrt{\frac{\tilde{D}_{KL}(\mathbf{x}_1, \mathbf{x}_2)}{2}}.$$

It is important to notice that (3.47) is not an upper bound on the KL divergence, but rather an approximation. This means that the above metric, i.e. $\vartheta'_{R1}(\mathbf{x}_1, \mathbf{x}_2)$ cannot provide strict security bounds, but still can provide approximate results whose evaluation is less computationally complex.

As a last remark, as suggested in [5, 6], let us highlight that G-OTS-C and G-OTS-R1 achieve similar security performance when the signals are sparse in the acquisition domain. In fact, in this case the security of the cryptosystem does not directly depend on k. Conversely, when signals are sparse in a frequency-like domain, e.g. in the DCT domain, it means that they are highly correlated in the acquisition domain. This implies that more information is leaked. As a matter of fact, because of the circulant structure of the sensing matrix, the more the signal correlation, the more the leaked information.

Furthermore, in [5], the authors show that if the signals are sparse in the DCT domain, then the distinguishability is independent of the signal size n. Conversely, if the signals are sparse in the acquisition domain, then the distinguishability decreases with n.

The G-OTS-R2 model we consider in the following will allow us to encompass this problem.

Confidentiality of G-OTS-R2

For the last scenario we consider the case of double randomization. Randomness is added to the matrix $\mathbf{R}$ whose diagonal entries are now i.i.d. following a Rademacher distribution. In other words, the signs of the input signals $\mathbf{x}_1$,$\mathbf{x}_2$ are randomized.

Let us recall a result from [6] which upper bounds the ϑ-distinguishability of a cryptosystem defined as above.

Theorem 3.4 *A G-OTS-R2 cryptosystem is at least $\vartheta_{R2}(\mathbf{x}_1, \mathbf{x}_2)$-indistinguishable w.r.t. $\mathbf{x}_1$, $\mathbf{x}_2$, where*

$$\vartheta_{R2}(x_1, x_2) = \sqrt{\frac{1}{4N_P^2 N_R^2} \sum_{r1,r2} \sum_{s1,s2} \left[\log \frac{|\mathbf{C}_{2,r2,s2}|}{|\mathbf{C}_{1,r1,s1}|} + \mathrm{Tr}(\mathbf{C}_{2,r2,s2}^{-1} \mathbf{C}_{1,r1,s1}) \right] - \frac{m}{4}} \tag{3.48}$$

$\mathbf{C}_{h,r,s} = \mathbf{P}_r \mathbf{C}_{\mathbf{R}_s \mathbf{x}_h} \mathbf{P}_r^T$, *for* $h = 1, 2$, *and* $N_R = 2^n$ *denotes the number of possible sign randomization matrices.*

Proof The proof follows the same lines as the proof of Theorem 3.3, however differently from G-OTS-R1 we have that $w_{1,r} = w_{2,r} = N_P^{-1} N_R^{-1}$. □

As with the result for G-OTS-R1, the upper bound depends on all the possible matrices $\mathbf{P}_r$ and $\mathbf{R}_s$. This means that the computation of the exact upper bound value might become too expensive for even very small values of m and n.

Unfortunately, in this case it is not possible to follow the same approximate approach obtained in Eq. 3.47. As a matter of fact, we have $\mathbf{C}_h = \sum_{r=1}^{N_P} \sum_{s=1}^{N_R} \frac{1}{N_P N_R} \mathbf{C}_{h,r,s} = \mathcal{E}_{\mathbf{x}_h} I_m$, meaning that for equal-energy signals the approximated KL divergence is zero. Nevertheless, by using the convexity bound approach, an approximate security metric can be obtained as

$$\vartheta'_{R2}(\mathbf{x}_1, \mathbf{x}_2) = \sqrt{\frac{1}{2N_R^2} \sum_{s1,s2} \tilde{D}_{KL}(\mathbf{R}_{s1}\mathbf{x}_1, \mathbf{R}_{s2}\mathbf{x}_2)}.$$

Again, the exact computation of the above metric may become too expensive for large values of n. In those cases, we can resort to Monte Carlo integration.

At this point it is important to underline that, as pointed out in [6], the double randomization offers the highest secrecy among the models we considered in this section. The G-OTS-R2 model provides a security which is independent of the sparsity k of the acquired signals and of the sparsity domain. Moreover, the distinguishability always decreases for large values of n regardless of the sparsity domain.

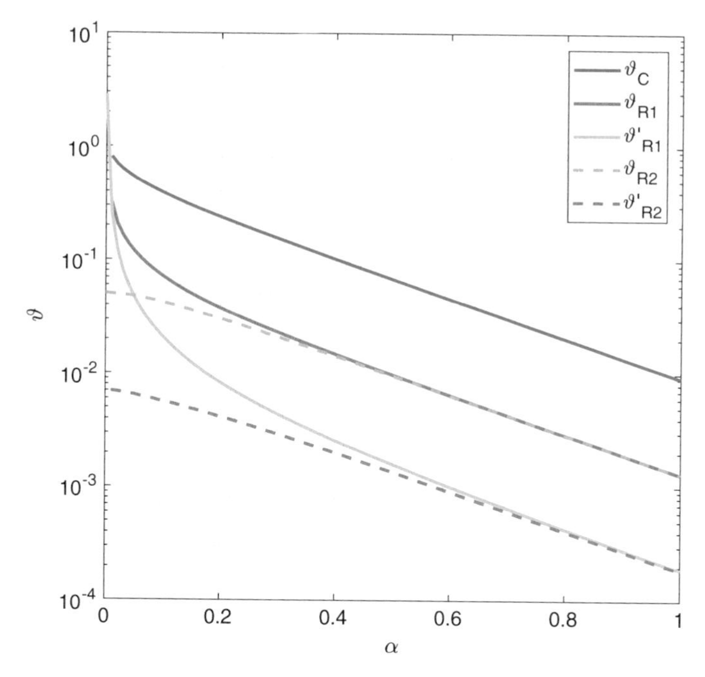

Fig. 3.7 Secrecy behavior of different implementations of circulant sensing matrices. The values of ϑ_C, ϑ_{R1}, ϑ'_{R1}, ϑ_{R2}, and ϑ'_{R2} are plotted for $\mathbf{x}_1 = [1, 0, \ldots, 0]$ and $[\mathbf{x}_2]_i = H(\alpha)e^{-4\alpha(i-1)}$ for $i = 1, \ldots, n$, where $H(\alpha)$ is a suitable value to normalize the energy of the signal

This confirms that this latter model, while more difficult to be numerically characterized and more expensive in terms of amount of randomness which needs to be generated at each encryption, achieves the best performances among all the considered models. In Fig. 3.7, we show an example of the behavior of the distinguishability for different circulant matrix randomizations. We set $\mathbf{x}_1 = [1, 0, \ldots, 0]$ and $[\mathbf{x}_2]_i = H(\alpha)e^{-4\alpha(i-1)}$ for $i = 1, \ldots, n$, where $H(\alpha)$ is a suitable value to normalize the energy of the signal. With this setting, a smaller distinguishability is expected for larger values of α. It is evident that the randomization can reduce the distinguishability of the signals. Moroever, while G-OTS-R1 is not very effective in hiding signals with very different structures, G-OTS-R2 can provide confidentiality irrespective of the underlying signal.

Let us conclude this section by considering the normalization strategy described in (3.6) for the case of circulant sensing matrices. If the matrices are i.i.d. Gaussian with no structure, then the aforementioned strategy allows perfect secrecy to be achieved. However, when the circulant structure is considered we can refer to the following theorem. Let us define $\mathbf{u}_{\mathbf{x}_h} = \mathbf{x}_h/\sqrt{\mathcal{E}_{\mathbf{x}_h}}$ and $\mathbf{u}_{\mathbf{y}_h} = \mathbf{y}_h/\sqrt{\mathcal{E}_{\mathbf{y}_h}}$, where $\mathbf{y}_h = \mathbf{\Phi}\mathbf{x}_h$, $h = 1, 2$. Then we have the following

Theorem 3.5 *The upper bounds given in* (3.42), (3.44), *and* (3.48) *computed for equal-energy signals* $\mathbf{u}_{x_1}, \mathbf{u}_{x_2}$ *holds also in the case of normalized measurements of generic signals* $\mathbf{x}_1, \mathbf{x}_2$.

Proof Let us define $\mathbf{y}'_i = \boldsymbol{\Phi}\mathbf{u}_{\mathbf{x}_i}$. It is easy to verify that $\mathbf{u}_{\mathbf{y}'_i} = \mathbf{y}'_i / \sqrt{\mathcal{E}_{\mathbf{y}'_i}} = \mathbf{u}_{\mathbf{y}_i}$. Then, we have the following inequalities involving the KL divergence

$$
\begin{aligned}
D_{KL}(\mathbf{y}'_1||\mathbf{y}'_2) =& D_{KL}(\mathbb{P}(\mathbf{u}_{\mathbf{y}_1}, \mathcal{E}_{\mathbf{y}'_1})||\mathbb{P}(\mathbf{u}_{\mathbf{y}_2}, \mathcal{E}_{\mathbf{y}'_2})) \\
=& D_{KL}(\mathbf{u}_{\mathbf{y}_1}||\mathbf{u}_{\mathbf{y}_2}) + D_{KL}(\mathbb{P}(\mathcal{E}_{\mathbf{y}'_1}|\mathbf{u}_{\mathbf{y}_1})||\mathbb{P}(\mathcal{E}_{\mathbf{y}'_2}|\mathbf{u}_{\mathbf{y}_2})) \\
\geq& D_{KL}(\mathbf{u}_{\mathbf{y}_1}||\mathbf{u}_{\mathbf{y}_2})
\end{aligned}
\tag{3.49}
$$

where we exploited the chain rule for KL divergence [10] and the fact that KL divergence is always nonnegative. Hence, the proof follows from the following chain of inequalities

$$
\delta(\mathbb{P}(\mathbf{u}_{\mathbf{y}_1}), \mathbb{P}(\mathbf{u}_{\mathbf{y}_2})) \leq \sqrt{\frac{1}{2} D_{KL}(\mathbf{u}_{\mathbf{y}_1}||\mathbf{u}_{\mathbf{y}_2})} \leq \sqrt{\frac{1}{2} D_{KL}(\mathbf{y}'_1||\mathbf{y}'_2)} \tag{3.50}
$$

where it is easy to verify that the right hand side of (3.50) evaluates to the upper bound on the distinguishability of equal energy signals. □

Lastly, even though out of scope of this book, we highlight that in [9] the distinguishability under the OTS-C scenario with $\boldsymbol{\phi}$ made of Bernoulli distributed entries (B-OTS-C) and ciphertext transmitted over a wireless channel is discussed. Similarly to the setting we addressed in Sect. 3.3.2, low channel gains and low PNR can improve the secrecy by reducing the KL divergence between distributions.

3.3.4 Upper Bound Validation

As previously done for G-OTS model, here we assess the quality of bounds discussed in this section and their behavior as a function of the related parameters. The attack model we consider in this case is similar to the ϑ-distinguishability experiment: an attacker has access to a measurement vector $\mathbf{y}$ and has to decide whether it is the result of the encryption of one of two equal energy signals $\mathbf{x}_1$ or $\mathbf{x}_2$. Then, the probability of detection and false alarm of the detector employed by the attacker are used as a performance metric. Based on its definition, the value of ϑ is obtained as the value which upper bounds $P_d - P_f$ for any possible detector $\mathcal{D}(\mathbf{y})$. Thus, it becomes important to consider the *best* possible detector. For a given probability of false alarm $P_f = \alpha$, by Neyman-Pearson (NP) lemma, the the detector which maximizes the probability of detection P_d is the one which sets $\mathcal{D}(\mathbf{y}) = \mathbf{x}_1$ whenever

$$
\Lambda(\mathbf{y}) = \frac{\mathbb{P}(\mathbf{y}|\mathbf{x}_1)}{\mathbb{P}(\mathbf{y}|\mathbf{x}_2)} \geq \tau \tag{3.51}
$$

where τ satisfies $\Pr\{\Lambda(\mathbf{y}) \geq \tau | \mathbf{x}_2\} = \alpha$.

If the sensing matrix entries are i.i.d., the measurements entries will be i.i.d. as well, and thus we can rewrite the NP test as

$$\Lambda'(\mathbf{y}) = \sum_{i=1}^{m} (\log(\mathbb{P}([\mathbf{y}]_i|\mathbf{x}_1)) - \log(\mathbb{P}([\mathbf{y}]_i|\mathbf{x}_2))) \geq \log \tau. \quad (3.52)$$

In the case of generic OTS model, a closed-form result for the NP detector is often difficult to obtain. Hence, the optimal NP test can be computed by substituting in the above equation the numerical approximations of $\mathbb{P}([\mathbf{y}]_i|\mathbf{x}_1)$ and $\mathbb{P}([\mathbf{y}]_i|\mathbf{x}_2)$ obtained according to (3.28).

Conversely, when considering circulant matrices with Gaussian entries, it is possible to directly evaluate the conditional measurements distributions and, as shown in [5], the optimal NP test in the case of G-OTS-C can be obtained as

$$\Lambda_C(\mathbf{y}) = \mathbf{y}^T(\mathbf{C}_2^{-1} - \mathbf{C}_1^{-1})\mathbf{y} \geq \tau'. \quad (3.53)$$

where $\tau' = \log \tau + \frac{1}{2}\log|2\pi\mathbf{C}_1| - \frac{1}{2}\log|2\pi\mathbf{C}_2|$.

Considering the G-OTS-R1 model, as pointed in [4], the optimal NP test can be computed as the ratio of two mixture distributions which, unfortunately, have an extremely high number of components even for relatively small values of m and n. An approximate solution, as discussed above for the practical evaluation of the KL divergence, is to approximate the mixture distributions as multivariate distribution with same mean and covariance matrix. This leads to the following test for G-OTS-R1

$$\begin{aligned}\Lambda_R(\mathbf{y}) = &\left(\frac{1}{\alpha_2} - \frac{1}{\alpha_1}\right)\mathbf{y}^T\mathbf{y} \\ &- \left(\frac{\beta_2}{\alpha_2(\alpha_2 + m\beta_2)} - \frac{\beta_1}{\alpha_1(\alpha_1 + m\beta_1)}\right)(\mathbb{1}^T\mathbf{y})^2 \\ &\geq \tau''\end{aligned} \quad (3.54)$$

where τ'' satisfies $\mathbb{P}\{\Lambda_R(\mathbf{y}) \geq \tau''|\mathbf{x}_2\} = P_f$. Let us note that this test cannot distinguish equal-energy signals whose components sum up to the same value in magnitude, i.e., such that $|\mathbb{1}^T\mathbf{x}_1| = |\mathbb{1}^T\mathbf{x}_2|$, since in this case we have $\alpha_1 = \alpha_2$ and $\beta_1 = \beta_2$. The above test is also ineffective in the case of equal-energy signals sensed with the G-OTS-R2 cryptosystem, since equal energy signals will yield measurements with the same covariance matrix.

Upper Bound Validation

Let us start by considering the case of generic OTS models. As done in Sect. 3.3.2, we consider $[\mathbf{x}_1]_i = 1/\sqrt{n}$ for $i = 1, \ldots, n$, and $[\mathbf{x}_2]_i = H(\alpha)e^{-\alpha(i-1)}$ for $i = 1, \ldots, n$, where $H(\alpha)$ is a suitable value to normalize the signal to unit energy. We show both the theoretical upper bound ϑ_{KL} and the result of the NP optimal test averaged over 10^6 independent realizations as a function of α in Fig. 3.8, considering a sensing matrix composed by i.i.d. uniform entries with zero mean and unit variance.

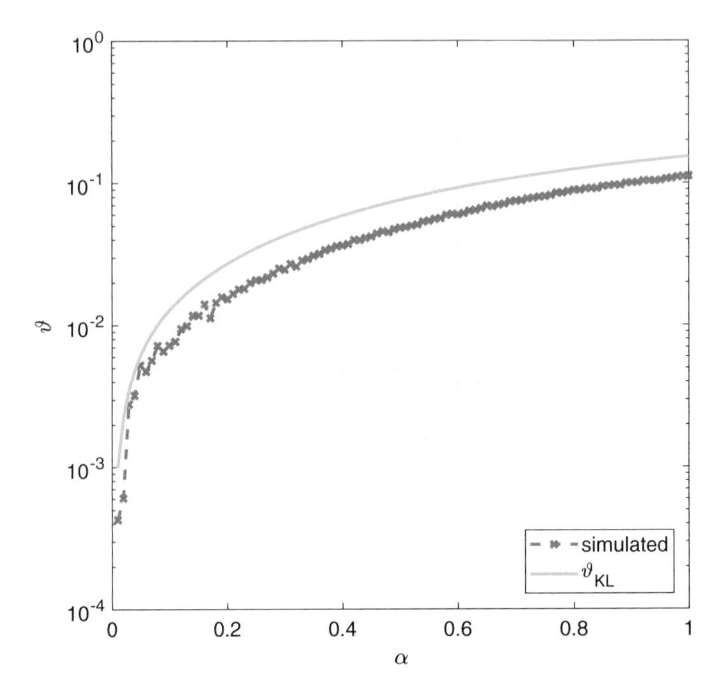

Fig. 3.8 Distinguishability of uniform energy signals under OTS model with uniform sensing matrix as a function of α

From the results in Fig. 3.8 it can be seen that the bound ϑ_{KL} is able to correctly predict the outcome of the simulated attack experiment, confirming that the distinguishability decreases as α approaches zero.

As a next step, in order to validate the bounds for structured sensing matrices, namely circulant matrices as of G-OTS-C and G-OTS-R1, we follow a similar approach. As done in Sect. 3.3.3, we set $\mathbf{x}_1 = [1, 0, \ldots, 0]$ and $[\mathbf{x}_2]_i = H(\alpha)e^{-4\alpha(i-1)}$ for $i = 1, \ldots, n$, where $H(\alpha)$ is a suitable value to normalize the signal to unit energy.

In Fig. 3.9, we show ϑ_C, ϑ_{R1}, ϑ'_{R1} and the result of the tests Λ_C and Λ_R averaged over 10^6 independent realizations, as a function of α. As for the previous experiments, we can confirm that the obtained bounds are able to well predict the behavior of an attacker trying to distinguish the two encrypted signals. Moreover, we can appreciate that the approximate bound ϑ'_{R1} is quite close to the performance achieved by the attack.

Average Case Behavior

We assessed that the theoretical bounds on the ϑ-distinguishability are able to predict with reasonable precision the performance of the specific attack we introduced at the beginning of this section. Hence, we will now consider the behavior of the different scenarios in the average case, when the sparsity is both in the acquisition domain and in the DCT domain, as a function of n and k. In order to provide these results, we will

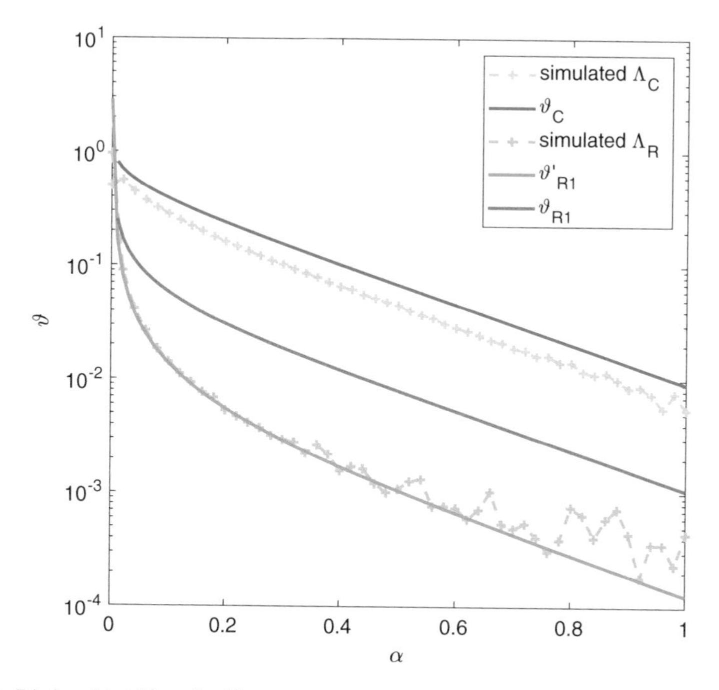

Fig. 3.9 Distinguishability of uniform energy signals under G-OTS-C and G-OTS-R1 models as a function of α. In case of G-OTS-R1 model both theoretical and approximated upper bounds are depicted

employ numerical upper-bounds which are obtained by computing ϑ_{KL} values over 1000 signal realizations and displaying the 0.95 percentile. The results are depicted in Figs. 3.10 and 3.11 for the case of generic sensing and in Figs. 3.12 and 3.13 for the G-OTS-C and G-OTS-R1 models.

In more detail, regarding the case of generic sensing matrices, in Figs. 3.10 and 3.11 we show the behavior of uniformly and Bernoulli distributed sensing matrices. It can be noticed that, in general, protecting the measurements through uniformly distributed sensing matrices provides higher secrecy with respect to the case of Bernoulli distributed ones. Moreover, as depicted in Fig. 3.10 when the signal is sparse in the acquisition domain the distinguishability decreases with k. Conversely, if the signals are sparse in the DCT domain, then the distinguishability is not affected by the sparsity k.

A different result can be appreciated in Fig. 3.11 where it can be noticed that, if the signals are sparse in the acquisition domain, then the distinguishability is constant with respect to the signal size n. Nonetheless, higher values of n lead to a vanishing distinguishability for those signals sparse in the DCT domain. As a last remark, it is important to highlight that the decreasing behavior of ϑ, either as a function of k or n depending on the sparsity domain, follows a similar rate.

Lastly, in Figs. 3.12 and 3.13 we show the same experiments as above but in the case of G-OTS-C, G-OTS-R1, and G-OTS-R2 models. It is immediate to notice

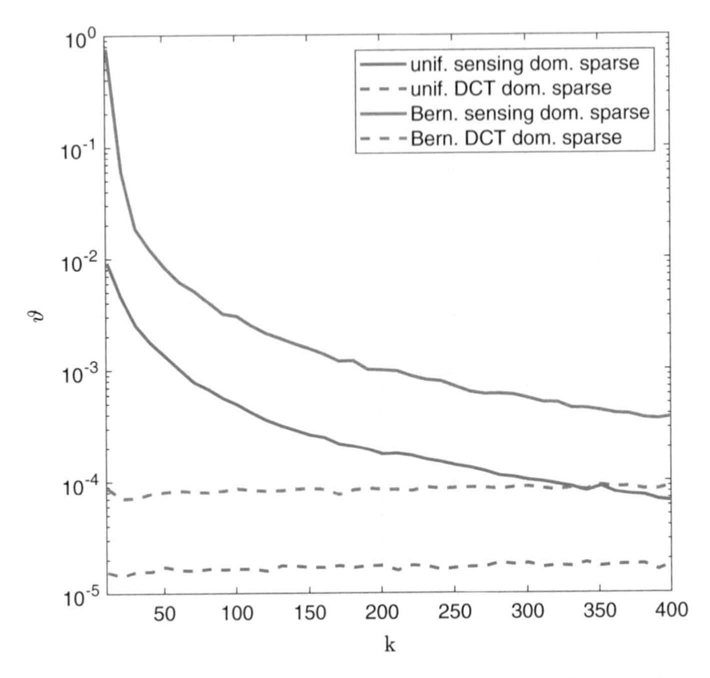

Fig. 3.10 Distinguishability of k-sparse signals for OTS model with uniform sensing matrices as a function of k. The sparsity is considered both in the sensing domain and in the DCT domain

that both G-OTS-R1 and G-OTS-R2 models, as expected, are able to provide higher security with respect to the standard G-OTS-C model. For G-OTS-C and G-OTS-R1 we have that the distinguishability is independent of k if the signal is sparse in the sensing domain, but it shows a strong relationship with k for signals that are sparse in the DCT domain. Indeed, as pointed in [5], sparsity in DCT domain implies high correlation in the original domain. In turn, since circulant sensing matrices do leak information about the correlation, then a strong dependence is expected. Interestingly, no dependency on k can be observed for G-OTS-R2, indicating that the sign randomization strategy is effective in hiding the signal structure.

If we then analyze the behavior as a function of the signal size n, we have that in case of G-OTS-C, ϑ decreases when n is large and the signal is sparse in the acquisition domain. Conversely it increases for large values of n in case of signals sparse in the DCT domain. Regarding the G-OTS-R1 model it is immediate to notice that, if the signals are sparse in the DCT domain, then ϑ is independent of n. Nevertheless, when the signals are sparse in the acquisition domain, ϑ significantly decreases. Finally, for the G-OTS-R2 model ϑ always decrease with n, irrespective of the sparsity domain.

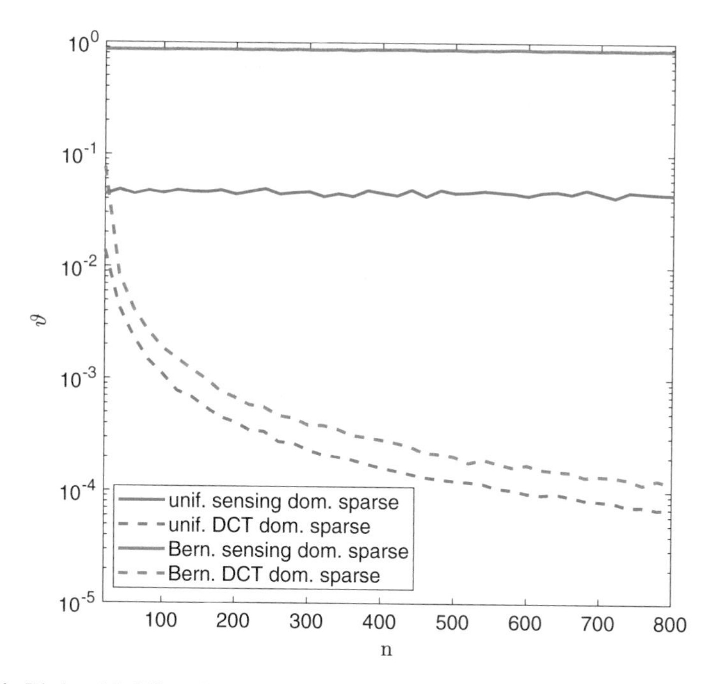

Fig. 3.11 Distinguishability of k-sparse signals for OTS model with uniform sensing matrices as a function of n. The sparsity is considered both in the sensing domain and in the DCT domain

3.4 Practical Sensing

3.4.1 *Overview*

Throughout this chapter we considered how both the structure and the probability distribution of the sensing matrix entries can affect the secrecy of a CS cryptosystem. Nevertheless, the above analysis did not take into account the fact that practical implementations require additional considerations. Though a complete discussion of the practical implementation of a CS cryptosystem is out of the scope of this book, we briefly discuss the main issues.

As a matter of fact, it is important to consider both the sensing matrix generation process, and the precision at which the entries can be represented. The key generation process should be computationally efficient since in the OTS model each encryption requires a fresh sensing matrix. In addition, it should make use of suitable cryptographic functions in order to ensure the required randomness.

Regarding the representation precision of the sensing matrix entries, it is important to recall that the above results on Gaussian i.i.d. sensing matrices are based on the infinite precision representation assumption. On the other hand, practical system can only deal with finite precision. An immediate consequence of the finite precision

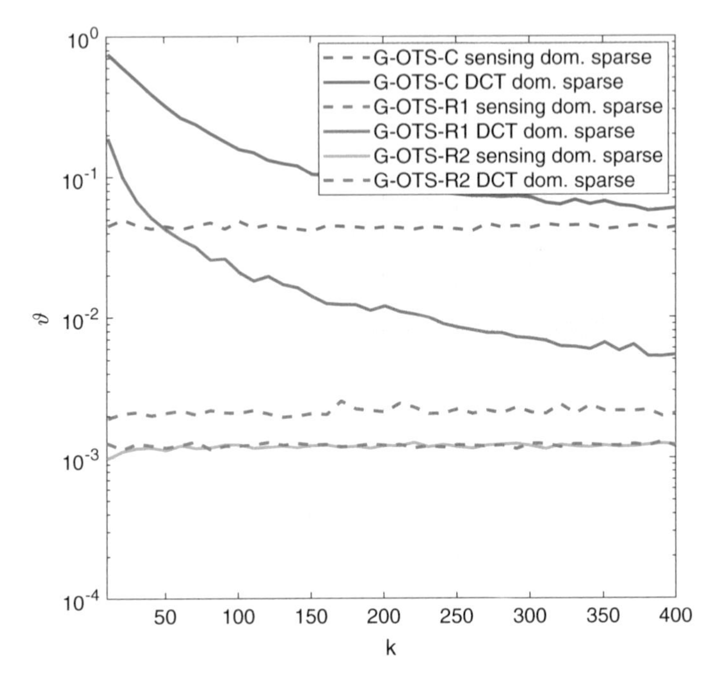

Fig. 3.12 Distinguishability of k-sparse signals under G-OTS-C, G-OTS-R1, and G-OTS-R2 models as a function of k. The sparsity is considered both in the sensing domain and in the DCT domain

representation is that a practical CS cryptosystem can offer only computational secrecy. Under a finite representation, there will be only a finite number of distinct sensing matrices. In principle, a computationally unbounded adversary could test all the possible sensing matrices an break the system. We recall here a result from the seminal work [22], stating that in the case of k-sparse signals, recovering the signal with a wrong sensing matrix will yield a dense vector with probability one. Hence, a brute force attack on the key space will succeed with probability one.

In the following we will give to the reader an overview of the two aforementioned issues occuring in practical CS cryptosystems.

3.4.2 Sensing Matrix Classes

As we previously hinted, because of the impossibility of using infinite precision representation in practical systems, not all the sensing matrix classes can be used as is. Sensing matrices with Bernoulli distributed entries are an example of sensing matrix class which can cope well with the finite precision requirement. However, the secrecy results for this kind of distribution only hold in asymptotic setting as shown in Sect. 3.3.2 and thus it is difficult to characterize the confidentiality of a

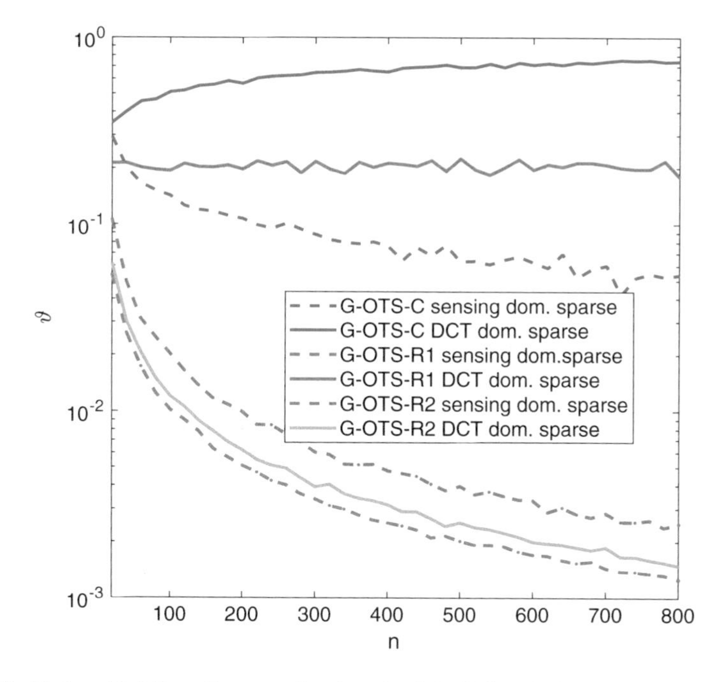

Fig. 3.13 Distinguishability of k-sparse signals under G-OTS-C, G-OTS-R1, and G-OTS-R2 models as a function of k. The sparsity is considered both in the sensing domain and in the DCT domain

given Bernoulli CS cryptosystem. Conversely, regarding the results on the secrecy provided by G-OTS model, we have shown that this specific model allows to reach high security performances, and in some cases, also perfect secrecy. However, the results we discussed are based on the assumption of being able to sample from Gaussian distributions with infinite precision, which unfortunately is not the case for practical systems.

The interesting problem of the secrecy of cryptosystems employing quantized Gaussian entries (QG-OTS) has been addressed by the seminal work in [27]. In more detail, the authors analyzed the experimental KL divergence between the measurements distributions as a function of the number of bits N_b over which the sensing matrix entries are quantized and value T_R at which the Gaussian distribution is truncated. The first and most important result discussed in [27] is that, as a general behavior, the KL divergence *exponentially* decreases as the number of quantization bits N_b increases leading to a conjectured relationship in the form of $\vartheta \propto 2^{-N_b}$. Further, since parameters N_b and T_R are shown to affect ϑ-distinguishability in a joint fashion, the approximated relationship between these parameters and ϑ-distinguishability is expected to be expressed as:

$$\vartheta \propto f(N_b, T_R, K) = \frac{\alpha T_R^{\beta}}{k 2^{N_b}}, \tag{3.55}$$

where α, β are hyper-parameters and k is the sparsity of the plaintext. Interestingly, the sparsity k can be put in relation with the asymptotical results in Sect. 3.2.4 for which, as the weighted sum of random variable increases, by the CLT different measurements will tend to the same distribution. Furthermore, the above function is increasing in T_R for fixed values of N_b and, is exponentially decreasing in N_b when T_R is fixed. This lead to the conjecture that must exist a regime condition for N_b and T_R, i.e. $T_R \propto \alpha_1 N_b^{\beta_1}$, for which ϑ exponentially decreases in N_b.

As a final remark, let us consider that, since we are considering a finite precision cryptosystem, we might also include a further quantization of the measurements. Simply put, because of the data processing inequality [10], additional operations performed on the ciphertext can decrease the mutual information between the resulting ciphertext and the plaintext.

An interesting case is when the spherical angle of measurements do not leak information about the spherical angle of the original signal as for G-OTS model, and the measurements are quantized to 1-bit. As a matter of fact, one-bit measurements will only depend on the spherical angle of the unquantized measurements which, under the G-OTS model, does not carry information. As a result perfect secrecy can be achieved. Interestingly, this setting has received a lot of attention from perspectives that do not necessarily consider the secrecy of the measurements. At first, it has been shown in [7, 18] that, if a suitable number of measurements is acquired, and the recovery algorithm enforces the consistency in the reconstruction, then is possible to recover the original signal within a scaling factor. Furthermore, the one-bit measurements setting, which has also been related to locality sensitive hashing in [1], has implications in the privacy-preserving embeddings which are discussed in detail in Chap. 4.

3.4.3 Sensing Matrix Generation

Here we briefly discuss a possible implementation strategy which can be used to generate entries following a Gaussian distribution. More specifically, we consider the sensing matrix having i.i.d. entries distributed according to $\mathcal{N}(0, 1)$. Thus, we need to securely generate mn Gaussian entries. Here we focus on the unstructured sensing matrices, but this discussion can be easily generalized to the case of structured sensing matrices. For example, if the sensing matrix has a circulant structure we only need to generate n entries.

At first, it is important to notice that transmitting at decryption side the full key (sensing matrix) would require to transmit mn entries. Understandably, this approach is unfeasible. For this reason, encryption and decryption modules should be kept synchronized by sharing a common seed, which is locally used to generate the full sensing matrix. Different approaches have been considered in order to account for this synchronization problem. As an example, in [13] the authors introduce the counter (CTR) mode of operation in the CS acquisition process, in analogy to the modes of operations commonly used in standard private-key cryptosystems. Given l plaintexts $\mathbf{x}_i$, for $i = 1 \dots l$, the generation of the sensing matrices $\mathbf{\Phi}_i$ for $i = 1 \dots l$ is controlled by a shared secret seed $\mathcal{K}$ and a counter CTR. More specifically, the counter is increased at every encryption, then sensing matrix entries are generated exploiting both the value of $\mathcal{K}$ and CTR by means of cryptographic hash functions. Lastly, the number of plaintexts to be encrypted must be smaller than the counter modulo, namely $l \leq 2^{n_M}$; if this condition is not met, it is possible to predict matrix reuse after 2^{n_M} encryptions.

We can now focus on the specific methods which can be used, given a shared secret seed $\mathcal{K}$ and the output of a finite state machine, such as a counter, to generate Gaussian samples. Different approaches to generate Gaussian entries can be envisioned, examples include the Box-Muller transform [24] and the Ziggurat method [21]. Because of its simplicity and efficient implementation, we might consider the Box-Muller transform method which generates two Gaussian distributed entries given two samples uniformly distributed in $(0, 1]$. To generate these uniform samples it is possible to rely on trusted cryptographic primitives which generate uniformly distributed random bits. In particular, as suggested in [14] let us consider the SHA3 *Skein* [15] or *Keccak* [3] algorithms which include a Key Derivation Mode which can generate an arbitrary number of uniformly distributed bits with a single call to the pseudorandom generation function (PRF). For their representation we use a fixed point representation which employs b-bits for the significand. If the b-bits are the output of a secure hashing algorithm, the result is a uniformly distributed number in $[0, 2^n - 1]$. Let us map this value in the interval $(0, 1]$, by dividing it by 2^n and assigning the 0 value to 1. The value of b directly influences the smallest number which can be represented in the above interval and hence the maximum output value of the Box-Muller transformation. As an example, if we consider $b = 32$ which leads to the smallest representable number to be 2^{-32}, the Box-Muller transformation will always output values smaller than 6.6604. Consequently, the Gaussian values are sampled, in practice, from a Gaussian distribution with truncated tails. In particular, by employing $b = 124$ the tails are truncated for probabilities smaller than 2^{-128} which is considered a negligible value for cryptographic applications.

Nevertheless, here we need to take into account that compressive cryptosystem schemes are suited for low-power weak-secrecy applications. This means that for this range of applications we are not interested in reaching perfect secrecy. Consequently, the number of required bits can be reduced depending on the specific application requirements in terms of secrecy and computational capabilities.

References

1. Andoni, A., Indyk, P.: Near-optimal hashing algorithms for approximate nearest neighbor in high dimensions. Commun. ACM **51**(1), 117–122 (2008)
2. Artstein, S., Ball, K.M., Barthe, F., Naor, A.: On the rate of convergence in the entropic central limit theorem. Probab. Theory Relat. Fields **129**(3), 381–390 (2004)
3. Bertoni, G., Daemen, J., Peeters, M., Van Assche, G.: The keccak sha-3 submission. Submission to NIST (Round 3) **6**(7), 16 (2011)

4. Bianchi, T., Bioglio, V., Magli, E.: Analysis of one-time random projections for privacy preserving compressed sensing. IEEE Trans. Inf. Forensics Secur. **11**(2), 313–327 (2016)
5. Bianchi, T., Magli, E.: Analysis of the security of compressed sensing with circulant matrices. In: 2014 IEEE International Workshop on Information Forensics and Security (WIFS), pp. 173–178 (2014)
6. Bianchi, T., Magli, E.: Security aspects of compressed sensing. In: Baldi, M., Tomasin, S. (eds.) Physical and Data-Link Security Techniques for Future Communication Systems, pp. 145–162. Springer International Publishing, Cham (2016)
7. Boufounos, P.T., Baraniuk, R.G.: 1-bit compressive sensing. In: 2008 42nd Annual Conference on Information Sciences and Systems (CISS 2008), pp. 16–21. IEEE (2008)
8. Cambareri, V., Mangia, M., Pareschi, F., Rovatti, R., Setti, G.: Low-complexity multiclass encryption by compressed sensing. arXiv:1307.3360 (2013)
9. Cho, W., Yu, N.Y.: Secure communications with asymptotically Gaussian compressed encryption. IEEE Signal Process. Lett. **25**(1), 80–84 (2018)
10. Cover, T.M., Thomas, J.A.: Elements of Information Theory. Wiley, New York (2006)
11. Do, M.N.: Fast approximation of Kullback-Leibler distance for dependence trees and hidden Markov models. IEEE Signal Process. Lett. **10**(4), 115–118 (2003)
12. Eldar, Y., Kutyniok, G.: Compressed Sensing: Theory and Applications. Cambridge University Press, Cambridge (2012)
13. Fay, R.: Introducing the counter mode of operation to compressed sensing based encryption. Inf. Process. Lett. **116**(4), 279–283 (2016)
14. Fay, R., Ruland, C.: Compressive sensing encryption modes and their security. In: 2016 11th International Conference for Internet Technology and Secured Transactions (ICITST), pp. 119–126. IEEE (2016)
15. Ferguson, N., Lucks, S., Schneier, B., Whiting, D., Bellare, M., Kohno, T., Callas, J., Walker, J.: The skein hash function family. Submission to NIST (round 3) **7**(7.5), 3 (2010)
16. Goldwasser, S., Micali, S.: Probabilistic encryption. J. Comput. Syst. Sci. **28**(2), 270–299 (1984)
17. Hershey, J., Olsen, P.: Approximating the Kullback Leibler divergence between Gaussian mixture models. In: ICASSP'07, vol. 4, pp. IV–317–IV–320 (2007)
18. Jacques, L., Laska, J.N., Boufounos, P.T., Baraniuk, R.G.: Robust 1-bit compressive sensing via binary stable embeddings of sparse vectors. IEEE Trans. Inf. Theory **59**(4), 2082–2102 (2013)
19. Kay, S.M.: Fundamentals of Statistical Signal Processing: Estimation Theory. Prentice-Hall, Upper Saddle River (1993)
20. LeCam, L.: Convergence of estimates under dimensionality restrictions. Ann. Stat. **1**(1), 38–53 (1973)
21. Marsaglia, G., Tsang, W.W.: The Ziggurat method for generating random variables. J. Stat. Softw. **5**(8), 1–7 (2000)
22. Rachlin, Y., Baron, D.: The secrecy of compressed sensing measurements. In: 2008 46th Annual Allerton Conference on Communication, Control, and Computing, pp. 813–817. IEEE (2008)
23. Schaad, J., Housley, R.: Advanced encryption standard (AES) key wrap algorithm (2002)
24. Scott, D.W.: Box-muller transformation. Wiley Interdiscip. Rev. Comput. Stat. **3**(2), 177–179 (2011)
25. Shannon, C.E.: Communication theory of secrecy systems. Bell Syst. Tech. J. **28**, 656–715 (1949)
26. Testa, M., Bianchi, T., Magli, E.: Energy obfuscation for compressive encryption and processing. In: 2017 IEEE Workshop on Information Forensics and Security (WIFS), pp. 1–6 (2017)
27. Testa, M., Bianchi, T., Magli, E.: On the secrecy of compressive cryptosystems under finite-precision representation of sensing matrices. In: 2018 IEEE International Symposium on Circuits and Systems (ISCAS), pp. 1–4 (2018)
28. Testa, M., Magli, E.: Compressive estimation and imaging based on autoregressive models. IEEE Trans. Image Process. **25**(11), 5077–5087 (2016)

29. Valsesia, D., Magli, E.: Compressive signal processing with circulant sensing matrices. In: 2014 IEEE International Conference on Acoustics, Speech and Signal Processing (ICASSP), pp. 1015–1019 (2014)
30. Yin, W., Morgan, S., Yang, J., Zhang, Y.: Practical compressive sensing with Toeplitz and circulant matrices. Visual Communications and Image Processing, vol. 7744, p. 77440K. International Society for Optics and Photonics, Bellingham (2010)
31. Yu, N.Y.: Indistinguishability and energy sensitivity of Gaussian and Bernoulli compressed encryption. IEEE Trans. Inf. Forensics Secur. **13**(7), 1722–1735 (2018)

Chapter 4
Privacy-Preserving Embeddings

Abstract In this chapter, we illustrate main results on privacy-preserving embeddings. Here, security properties of embeddings are analyzed by considering two possible scenarios for their use. In the first case, a client submits a query containing sensitive information to a server, which should respond to the query without gaining access to the private information. This is discussed describing an authentication system in which a client submit an embedding of a physical characteristic of a device, and a verification server is able to match the embedding without revealing the actual physical characteristic. Interestingly, in this case the security properties of the embedding permit to combine it with existing biometric template mechanisms, enhancing the security of the system. In the second case, a large amount of sensitive data is stored in the cloud and a user should be able to make specific queries to the cloud without gaining access to the data. Here, we describe a universal embedding that preserves distances only locally. If data are stored in the cloud using this embedding, a user is able to retrieve data close to the query, but the complete geometry of the dataset remains hidden by the embedding and data cannot be recovered.

Content-based retrieval, user authentication, recommender systems and many more are becoming widespread and used on a daily basis. Such systems deal with sensitive information and involve multiple parties which cannot be trusted and have often competing privacy goals. A basic scheme involves a client transmitting a query containing sensitive data to a server, that processes it in order to respond to the client with information from its database. There are at least two privacy goals in a scenario of this kind: from the client's perspective, the goal is to protect the sensitive information contained in the query; from the server's perspective, the goal is to protect the information contained in the database.

Many techniques used in such systems can be reduced to comparing signals in order to determine their similarity. Signals are treated as points in a metric space and a distance function, such as the Euclidean distance or the Hamming distance, measures their similarity. Signal embeddings, as introduced in Sect. 2.2, can provide privacy-preserving properties in many ways. This chapter is going to use the client-server perspective to analyze what kind of properties of embeddings can be leveraged to protect private information.

M. Testa et al., *Compressed Sensing for Privacy-Preserving Data Processing*,
SpringerBriefs in Signal Processing, https://doi.org/10.1007/978-981-13-2279-2_4

In the following sections we are going to discuss two examples in order to explain how embeddings and their properties can be used to build privacy-preserving systems. The first example is a user authentication system in which a user is recognized through a physical characteristic of their smartphone. This problem involves a mobile client with the physical characteristic which should be kept private, but still allowing a server to verify it. From the server's perspective, it must verify the validity of the characteristic without revealing it.

The second example is a data mining system performing clustering of data submitted by multiple users. The data are sensitive so the server should not be able to understand their content. The data form clusters and the inter-cluster distances are much larger than the intra-cluster distances. The goal of the server is to understand for each user which other users have submitted similar data, i.e., discover one of the clusters. The server should not be able to infer the full geometry of the data, e.g., the inter-cluster distances, in order to avoid it being able to reconstruct the data.

4.1 User Authentication

4.1.1 Overview

User authentication is an ever-growing concern due to the popularity of web-based services like banking, social networks, etc. [3]. However, the traditional scheme based on secret passwords is inadequate for modern times. People use a multitude of services and should adopt different passwords for each of them. Also, long and unpredictable passwords should be generated for security reasons but remembering them is challenging. As a consequence, short and easily predictable passwords are commonly reused, considerably reducing the security of the system.

Recently, multifactor authentication schemes have been proposed, where the knowledge factor, i.e., the secret password, is complemented with the possession of one, or more, physical or software tokens [7]. A novel scheme to verify a possession factor has been proposed in [27]. It relies on recognizing a smartphone through a physical property of its digital image sensor named photo-response non-uniformity (PRNU). The PRNU is a sensor-specific multiplicative noise pattern that has enjoyed great popularity in the last decade because it has several applications such as determining which camera has acquired a given photo [12, 20], camera-based image retrieval [25, 26] or clustering [8, 17] and detecting and localizing image forgeries [10, 19]. Nowadays smartphones are ubiquitous, so a system that exploits a characteristic of an object everyone has in their pockets is potentially very interesting. A system could be envisaged where the smartphone is used as key thanks to the unicity of the PRNU of its camera. However, turning this idea into a practical authentication system requires to solve several important problems, as well as rigorously show the security of such solutions.

First, the PRNU survives JPEG compression, as well as some image processing operations, and it can be found in photos that are publicly available, e.g., on social networks [13]. However, the PRNU is inevitably degraded by such operations in the form of losing or severely degrading its high-frequency content. On the other hand, in the framework of user authentication, the legitimate user has full control over the camera and could extract the PRNU with an arbitrarily high quality. Therefore, it is clear that high-frequency components of the PRNU extracted from RAW images are secret data that cannot be found in publicly available data, assuming that RAW images are never disclosed.

Furthermore, there are desirable privacy objectives at both the client and the server. The client should avoid revealing the PRNU to a server. On the other hand, the server should verify the correctness of the PRNU without explicitly storing it. In Sect. 4.1.3 a detailed security analysis is provided on the robustness of the system to attacks reading the information stored by the system on the client and on the server. Many of the results hinge on the use of signal embeddings.

A variant of sign random projections can be used to compute an embedding of the high-frequency PRNU components. This embedding has two main purposes: (i) reducing the dimensionality of the input to a low-dimensional binary space; (ii) providing a security layer allowing the actual PRNU to never be disclosed by treating the sensing matrix as a secret, and possibly revoking the compressed version by changing the random projection matrix. Since the server should not store a copy of the PRNU, or its compressed version, techniques used for biometric template protection [11, 18] are coupled with the embedding. Namely, an implementation of a secure sketch and a fuzzy extractor based on polar codes is used. The main idea is that during the registration phase a secure sketch is created by combining a uniformly random secret and the compressed fingerprint, and stored by the server: this sketch alone does not reveal anything about the compressed fingerprint. At verification time, an estimate of the compressed fingerprint is presented and combined with the sketch. If the Hamming distance between this estimate and the one used for registration is sufficiently low, the polar code can successfully correct all the bit errors and recover the uniform secret thus validating the user.

This implementation also addresses the fact that residual correlation might exist in the embedded space when an attacker only has JPEG images. However, the embedding ensures that there exists a gap between Hamming distances exhibited by valid PRNU patterns (low distance values) and distances exhibited by PRNU patterns estimated by the attacker. A specific construction uses a coding technique for the wiretap channel based on polar codes to effectively prevent the attacker from gaining access to the system.

4.1.2 System Description

The main idea is to use the PRNU fingerprint of the camera sensor of a user's device, e.g., a smartphone or a tablet, as a weak physical unclonable function [14] for authentication. An overview block diagram is shown in Fig. 4.1.

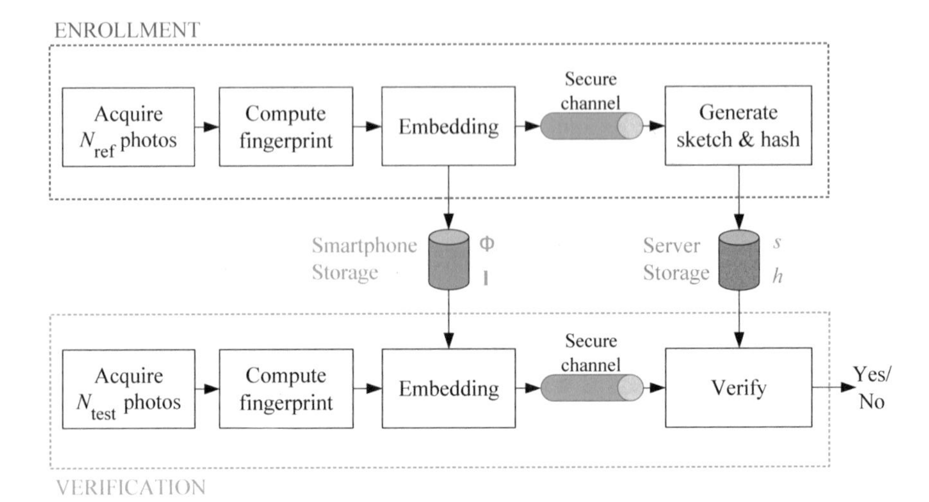

Fig. 4.1 System block diagram

In a first phase, the user enrolls into the system by providing a high quality estimate of the device fingerprint, obtained from a certain number of photos acquired in controlled conditions. This fingerprint is real-valued and has as many entries as the number of pixels in the sensor. The system compresses it by means of binary-quantized random projections and also stores some side information in the form of the seed of the pseudorandom number generator and the positions of the entries with largest magnitude (outliers) within those random projections before quantization which will be then used in the authentication phase. The exact algorithm as well as the role of the outliers will be made clear in the following sections. This side information is treated as a secret and accurately protected. At the server side, the compressed fingerprint is processed by a fuzzy extractor. Namely, the server generates a uniformly random bit string from the compressed fingerprint and stores a secure hash of this bit string, together with a secure sketch of the fingerprint.

In the authentication phase, the user reproduces a noisy version of the device fingerprint by acquiring a fresh set of photos and compressing the resulting fingerprint according to the stored side information. The server then uses the fuzzy extractor scheme for reproducing the secret bit string from the received compressed fingerprint and the secure sketch, and compares the recovered bit string with the stored secure hash. If the user provides a version of the compressed fingerprint sufficiently close to the enrolled one, then the server can reproduce the same bit string of the enrollment phase and grants access to the system; otherwise, it denies access.

Fingerprint Extraction

The high-frequency components of the PRNU pattern estimated from RAW images allow to estimate a fingerprint that is capable of discriminating different sensors and, at the same time, that is ideally uncorrelated with any estimate that can be extracted from JPEG images. Since the RAW acquisition process can be controlled and the fingerprint extraction has to run efficiently on a user's smartphone, an assumption is made that the user acquires approximately flat images.

The RAW image is first demosaiced and color calibrated to obtain image $\mathbf{o} = [\mathbf{r}, \mathbf{g}, \mathbf{b}]$. The luminance component of such image is then obtained by applying the transformation

$$\lambda = 0.299\mathbf{r} + 0.587\mathbf{g} + 0.114\mathbf{b}.$$

It is possible to extract an estimate of the high-frequency components of the PRNU pattern to be used as fingerprint by means of a highpass filter (hereafter denoted as HPF) applied to the luminance component of the demosaiced and color calibrated image. This filter can be implemented as a product in the DCT domain. When multiple images $\mathbf{o}^{(l)}$ are available the fingerprint is jointly estimated as

$$\mathbf{k}^{\mathsf{RAW}} = \frac{\sum_l \mathbf{o}^{(l)} \cdot \mathsf{HPF}\left(\lambda^{(l)}\right)}{\sum_l (\mathbf{o}^{(l)})^2}. \quad (4.1)$$

However, some non-unique artifacts (NUA) [9] may be present, e.g., because of color filter array (CFA) interpolation, sensor linear pattern, etc. Such artifacts may introduce ambiguities in the camera detection process and should be removed. Hence, as a post-processing operation, row and column means in a checkerboard pattern are removed and Wiener filtering is used to suppress any periodic artifacts.

Fingerprint Embedding

Since the fingerprint must be sent to a server for verification purposes, it is of paramount importance to compress it to a size that makes transmission over bandlimited channels manageable. The objective of the compression step is to transform the real-valued, high-dimensional fingerprint into a short binary code. Correlated fingerprints must be mapped into similar binary codes.

In Sect. 2.2 we presented binary-quantized random projections, characterized by the property that their Hamming distance concentrates around the angle between the original uncompressed fingerprints. One can therefore use them to obtain compact binary codes. Since the fingerprints are high-dimensional objects, a complexity issue arises in the calculation of the random projections. This can be solved by using circulant random matrices [2, 15, 22, 28, 30, 31] with randomized column signs, as shown in [25]. For such matrices, only the first row must be generated at random and the matrix-vector product can be efficiently performed using the FFT.

A modified version of such random projections, called adaptive embedding [29], as presented in Sect. 2.2, can be used to trade some randomness for a better (more compact) representation of signals correlated with a particular signal of interest. This solution has three main advantages in the context of the user authentication system:

- save transmission time;
- more efficient and easier design of the fuzzy extractor at server side;

- adaptivity allows to preserve as much as possible of the inter-class correlation gap between fingerprints extracted from JPEG data and fingerprints extracted from RAW data; this also simplifies the design of the channel code in the fuzzy extractor because it maximizes the margin between the bit-error probability observed by a legitimate user and that observer by an attacker.

During the registration phase, a high-quality estimate of the fingerprint $\mathbf{k} \in \mathbb{R}^n$ is available. A vector ϕ with n i.i.d. Gaussian entries is generated and circularly convolved with $\mathbf{k}$ using the FFT to implement a circulant sensing matrix. The result of this operation is first subsampled to keep the first m_{pool} values. The $m < m_{\text{pool}}$ entries with largest magnitude are identified and their locations $\mathbf{l}$ stored locally on the user device as side information. Finally, the sign of the entries at those locations is saved as compressed fingerprint w of m bits. During the verification phase, a test fingerprint $\mathbf{k}'$ is presented for compression, and its projections are computed by keeping only the sign of the entries indexed by $\mathbf{l}$.

The value of m_{pool} determines the storage overhead required for the location information. Choosing m outliers from a larger pool improves the adaptivity to the reference signal but increases the storage overhead. The effect of adaptivity is shown in Fig. 4.2 where the expected value of the Hamming distance between the binary codes is plotted against the correlation coefficient between the original uncompressed fingerprints. Notice that the adaptive method allows to achieve smaller values for the Hamming distance and maximize the margin between the class of invalid fingerprints having very low correlation values and the class of valid fingerprints having higher correlation values. This is shown in Figs. 4.3 and 4.4 which report the results of an experiment on 14 devices. It can be seen that there exists a gap between the values of correlation coefficient ρ when computed between an uncompressed fingerprint

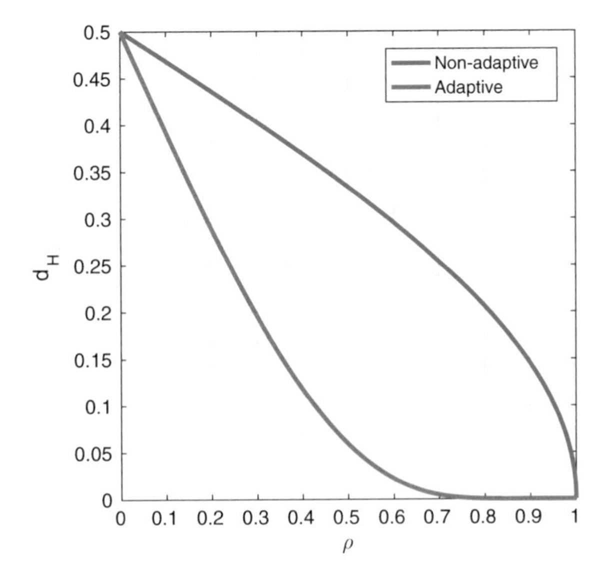

Fig. 4.2 Expected value of Hamming distance in the embedded space against correlation coefficient in the original space for adaptive and non-adaptive sign random projections. $m_{\text{pool}} = 2^{20}$, $m = 2^{15}$

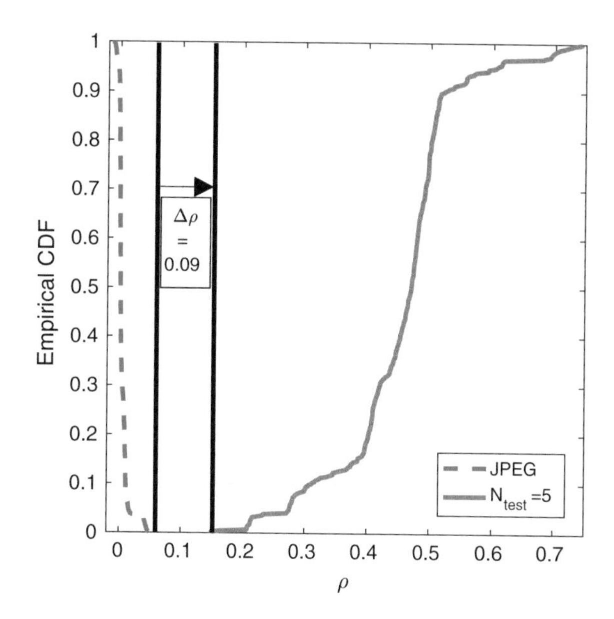

Fig. 4.3 Empirical cumulative distribution for the experimentally measured correlation coefficients before fingerprint compression. Reference fingerprint is from 20 RAW images. Test fingerprint is either from 200 JPEG images or from 5 RAW images. For JPEG the complementary CDF is plotted for visualization purposes

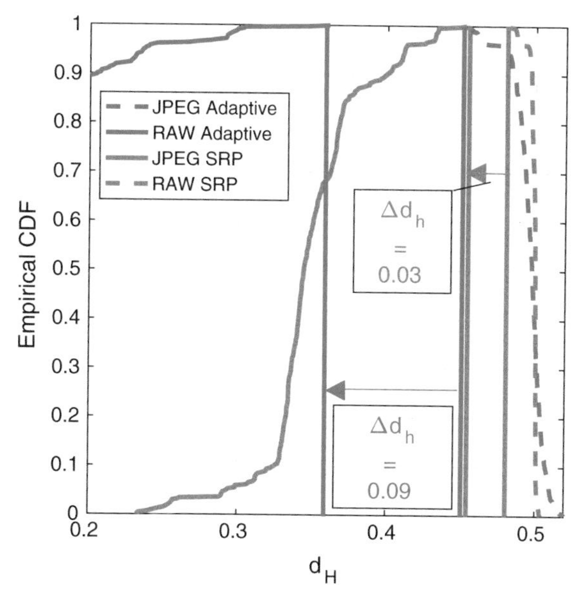

Fig. 4.4 Empirical cumulative distribution for the experimentally measured Hamming distances after fingerprint compression with adaptive embedding or sign random projections (SRP). Reference fingerprint is from 20 RAW images. Test fingerprint is either from 200 JPEG images or from 5 RAW images. For JPEG the complementary CDF is plotted for visualization purposes. $m = 2^{15}$, $m_{pool} = 2^{20}$

extracted from RAW images and an uncompressed fingerprint extracted from JPEG images or an uncompressed fingerprint also extracted from RAW images. The adaptive embedding allow keeping this gap as wide as possible when measured in terms of Hamming distances after compression.

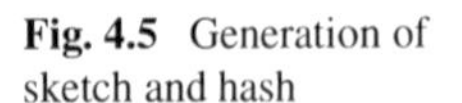

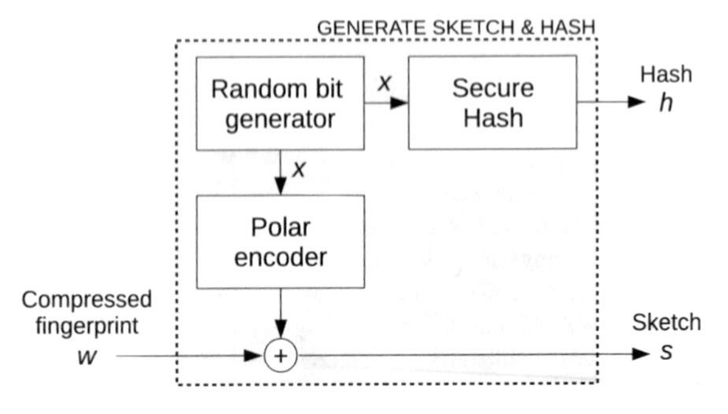

Fig. 4.5 Generation of sketch and hash

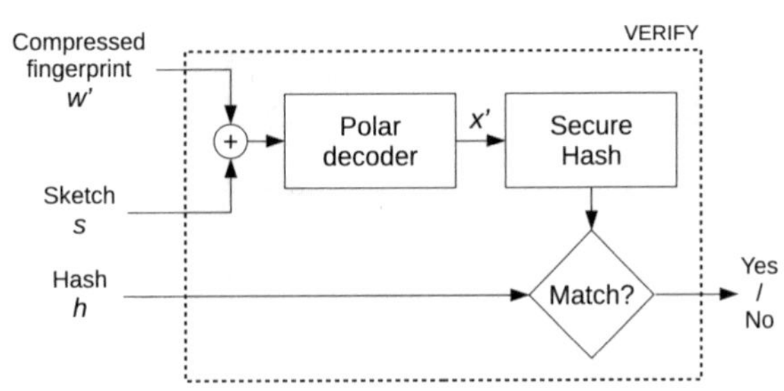

Fig. 4.6 User verification

User Registration and Verification

Due to the non-exact repeatability of the PRNU fingerprint estimation procedure, during the verification phase the user will produce a compressed fingerprint that contains some bit errors with respect to the enrolled fingerprint. Moreover, an attacker having access to a certain number of publicly available JPEG photos acquired by the user's device may also be able to provide a noisy version of the enrolled fingerprint, albeit with a much higher number of bit errors.

In order to cope with this scenario, a fuzzy extractor scheme based on channel codes [16] and a coding scheme for the wiretap channel that uses polar codes [21] are used. The scheme can be summarized by the generation function used during registration, depicted in Fig. 4.5 and the verification function, depicted in Fig. 4.6.

During the enrollment phase, the server generates a uniformly random string x of k bits. From this secret string, the server computes a hash $h = \mathsf{SH}(x)$, where $\mathsf{SH}(\cdot)$ denotes a secure hashing function, and a secure sketch $s = w \oplus C(x)$, where w is the compressed fingerprint received from the user and C denotes a (m, k) error correcting code based on polar codes. The server then discards x and stores h and s.

During the verification phase, the server computes the k-bit string $x' = D(w' \oplus s)$, where w' is the noisy fingerprint and D denotes the decoding algorithm of the error correcting code, and authenticates the user only if $\mathsf{SH}(x') = h$.

The error correcting code is not a standard (m, k) polar code, but is constructed according to the scheme in [21], considering two binary symmetric channels, one for the legitimate user with bit flip probability p_l determined by the Hamming distance between compressed fingerprints of the same sensor; the other for an attacker with

JPEG images with bit flip probability $p_a > p_l$. According to Fig. 4.4, experimentally determined values could be $p_l = 0.40$ and $p_a = 0.45$. Notice that, as a safety margin, $p_l = 0.40$ is above the highest Hamming distance observed in the experiments with matching RAW fingerprints and $p_a = 0.45$ is below the lowest value observed with JPEG fingerprints. Therefore, the design of the error correcting code should put the phase transition between high and low probability of error correction in between 40 and 45% bit errors. Notice how this design is helped by the adaptive embedding. Figure 4.4 also reports the values that would have been used for an embedding based on sign random projections, which are $p_l = 0.45$ and $p_a = 0.48$. This would require a code with a higher rate in order to correct more errors but also a code with a longer block size in order to have a sharp enough phase transition to fit the smaller gap.

4.1.3 Security Analysis

The security of the proposed authentication scheme depends on the probability that an attacker gains access to the system. Two important security assumptions are made: (i) the attacker does not have access to any RAW photos of the user's device; (ii) the attacker can access only a finite number N of JPEG photos of the user's device.

Four different security scenarios can be considered, depending on whether the attacker can access the parameters stored on the server, i.e., the sketch s and the secure hash h, or the parameters stored by the client, i.e., the seed for generating $\mathbf{\Phi}$ and the location vector $\mathbf{l}$. All the scenarios are carefully described in [27]. In the following we will focus on two scenarios that hinge on the properties of the embedding, i.e., when the information at the client is not compromised.

In order to formally analyze the security of the proposed system, it is useful to characterize the secrecy of a variable according to the probability of an adversary guessing its value. If the adversary has no observations about A, the best strategy is to choose as guess the most likely value. Formally, the facility with which an adversary guesses A can be measured by the min-entropy of A, defined as

$$H_\infty(A) = -\log(\max_a \Pr[A = a]).$$

Min-entropy can be considered a worst-case entropy. A variable whose min-entropy is m bits is as hard to guess as a uniformly random string of m bits.

If the adversary observes a variable B which is in some way correlated with A, the probability of guessing A usually increases. This can be formally expressed by the average min-entropy of A given B, defined as

$$\tilde{H}_\infty(A|B) = -\log(E_b[2^{H_\infty(A|B=b)}]).$$

The rationale of the above definition is that, given an observable variable B on which the adversary has no control, the advantage of the adversary can be measured by the expected value of the probability of guessing A over the distribution of B.

Scenario 1: Server and Client are Both Safe

This is the best case scenario, when the attacker can only access N JPEG photos. Let us denote these photos as $\mathbf{O}_a$. It is easy to see that these photos give no information about the enrolled compressed fingerprint w:

Lemma 4.1 *If $\mathbf{\Phi}$ is made of i.i.d Gaussian variables and $w = \text{sign}(\mathbf{\Phi}_\mathbf{l}\mathbf{k})$, where $\mathbf{\Phi}_\mathbf{l}$ denotes the submatrix of $\mathbf{\Phi}$ formed by the rows indexed by $\mathbf{l}$, then $\widetilde{H}_\infty(W|\mathbf{O}_a) = m$.*

Proof We note that $I(W;\mathbf{O}_a) \leq I(W;\mathbf{k}) = 0$. The first inequality holds since $\mathbf{O}_a \rightarrow \mathbf{k} \rightarrow W$ is a Markov chain, whereas the last equality is due to the fact that W depends only on the spherical angle of $\mathbf{\Phi k}$ that, for $\mathbf{\Phi}$ drawn from a Gaussian ensemble, is independent from $\mathbf{k}$ and uniformly distributed on the unit sphere (see see Lemma 3.1 and Lemma 1 in [1]). Hence, $\widetilde{H}_\infty(W|\mathbf{O}_a) = H_\infty(W) = m$.

In this scenario, the best an attacker can do is to draw w uniformly at random. The probability of success of this attack can be upper bounded as follows:

Theorem 4.1 *Under Scenario 1, the probability of success of the attacker verifies $P_a \leq 2^{-k}$.*

Proof Under a fixed x, the attacker succeeds if he/she chooses $w \in \mathcal{C}_x$, where $\mathcal{C}_x = \{w|\mathsf{SCD}_k(w) = x\}$ and $\mathsf{SCD}_k(\cdot)$ denotes the output of the successive cancellation decoder for polar codes on the k bits corresponding to x. Since x is uniformly random and not under the attacker's control, the probability of success is obtained as

$$P_a = \mathbb{E}_x\left[\mathbb{P}\left(w \in \mathcal{C}_x\right)\right] = \mathbb{E}_x\left[\frac{|\mathcal{C}_x|}{2^m}\right] = \frac{1}{2^{m+k}}\sum_x |\mathcal{C}_x| \leq \frac{1}{2^k}$$

where the last inequality holds since $\mathcal{C}_x$ are disjoint sets.

Lemma 4.1 relies on the statistical properties of random projections, which were extensively analyzed in Sect. 3.1. A possible remark on the above results is that Lemma 4.1 does not hold if $\mathbf{\Phi}$ is circulant, since measurements obtained according to a circulant matrix do not exhibit perfect spherical secrecy. In such cases, we can have the same result by modifying fingerprint compression as $w = \text{sign}(\mathbf{\Phi}_\mathbf{l}\mathbf{k}) \oplus \mathbf{b}$, where $\mathbf{b}$ is a uniformly random m-bit vector that the client stores along with $\mathbf{\Phi}$ and $\mathbf{l}$. Alternatively, we can exploit the results of Sect. 3.3.3 and apply a proper randomization to $\mathbf{\Phi}$.

Notice that in this scenario the security is guaranteed by the secrecy of the projection matrix and an attacker may even have access to RAW photos without compromising the system.

Scenario 2: Server is Compromised

Under this scenario, the attacker can see $\mathbf{O}_a$, s, and h. Let us consider an intermediate scenario in which the attacker observes only s. Since $H_\infty(W) = m$, due to the properties of the fuzzy extractor the secret x is indistinguishable from a uniformly random k-bit vector, even when observing s. This can be equivalently stated as $\tilde{H}_\infty(W|S) = k$, i.e., s is a (m, m, k, p_l, α)-secure sketch (this result can be proved using Lemma 4.5 in [11]).

Thanks to Lemma 4.1, the above result holds also when the attacker observes $\mathbf{O}_a$ and s, since $\tilde{H}_\infty(W|S, \mathbf{O}_a) = \tilde{H}_\infty(W|S)$. This follows from the fact that $\mathbf{O}_a \to W \to S$ is a Markov chain and W is independent from $\mathbf{O}_a$. In both scenarios, the attacker can only guess x', pick a random r, and try whether $w' = C(x', r) \oplus s$ is accepted by the system, which has a success probability $P_a = 2^{-k}$.

When the attacker also observes h, the system does not satisfy any more the above statistical security definition, since it is easy to verify $\tilde{H}_\infty(W|S, H = h) = \log N_C(h)$, where $N_C(h)$ is the number of collisions yielding value h in the secure hashing function, when computed over all possible x. In this scenario, the attacker is able to verify any feasible w until he/she finds a w' satisfying $w' = s \oplus C(x', r)$, with $\mathsf{SH}(x') = h$. Nevertheless, with a proper secure hashing function the system is computationally secure:

Proposition 4.1 *If* $\mathsf{SH}(\cdot)$ *is ideal, then under Scenario 2 the expected complexity of an attack is* $N_a = \Omega\left(2^{\min\{k, n_{hash}\}}\right)$ *operations, where* n_{hash} *is the hash length in bits.*

Proof If $k > n_{\text{hash}}$, a pre-image attack on a n_{hash}-bit ideal secure hash requires $2^{n_{\text{hash}}}$ guesses on average. If $k < n_{\text{hash}}$, finding the right x requires $\frac{2^k+1}{2}$ guesses on average.

Again, the security of the scheme comes from the properties of the secure embedding that transforms an arbitrary signal into a uniformly distributed bit string. Without the proposed embedding, the fuzzy extractor would not be able to protect all the k secret bits. Hence, it is evident how secure embeddings can be combined with other cryptographic primitives in order to enhance their privacy.

4.2 Bounded-Distance Clustering

4.2.1 Overview

Previous sections and chapters have shown how embeddings, especially when constructed from random projections, are an important tool for information processing. This chapter presents recent results [4, 6] where a particular construction, called *universal embedding*, allows to achieve embeddings which are more general in terms of the function distorting the geometry of the signal set. This is particularly interesting because it creates embeddings with specific properties. An example is introducing a distortion in the geometry of the signal set such that only a range of distances is accurately preserved while all information about the mutual relationships between points lying at distances outside this range is lost.

A careful reader would notice that we have already presented an embedding distorting signal distances, in the form of the adaptive embedding used in the user authentication problem of Sect. 4.1.2. The universal embeddings that we are going to review in the next section essentially differ by the fact that they are non-adaptive, i.e., they do not rely on training or reference signals to optimize the embedding. Rather they always create the same distortion of the signal space regardless of the input signals. A more detailed comparison between the adaptive and universal embeddings is presented at the end of Sect. 4.2.2.

While universal embeddings are a quite general construction that allows design choices in terms of the distortion on signal distance and in terms of the rate of the representation, i.e., the number of bits used to represent each value after the embedding, an interesting use case is when a binary embedding is created. This binary universal embedding can be seen as a generalization of sign random projections, due to the different quantization function, that allows to provide a bound on the maximum distance that is accurately represented by the embedding. It also possible to prove that this construction has a interesting privacy-preserving property due to the fact that the mutual information between the binary embeddings of two signals lying beyond the maximum distance quickly goes to zero. This means that it is impossible to infer the full geometry of the original signal set from the embeddings alone. We are going to present an application of this in Sect. 4.2.3 to clustering signals forming communities, i.e., with low distances inside the community and large distances between communities.

4.2.2 Universal Embeddings

Universal embeddings were first introduced in [4, 5], in the form of scalar quantized random projections with a special construction of the quantizer, called a universal scalar quantizer, being a periodic function with uniform quantization bins. In particular, the latter work addressed the problem from a compressed sensing perspective showing that the uniform scalar quantizer allows to achieve rate-efficient representations, i.e., with exponentially-decreasing distortion on the reconstructed signal as a function of the number of measurements. Later works such as [23, 24] expanded on that showing rate-distortion gains with practical reconstruction schemes and a vector extension of the universal scalar quantizer. Finally, the method was generalized in [6] for periodic scalar quantizers of arbitrary shape and random projection matrices of arbitrary distribution.

A universal embedding of signal $\mathbf{u}$ is computed by the following operation

$$\mathbf{y} = f(\mathbf{u}) = h\left(\mathbf{\Phi u} + \mathbf{w}\right) \tag{4.2}$$

where $\mathbf{\Phi}$ is an $m \times n$ random matrix with i.i.d. entries, $\mathbf{w}$ a dither noise with i.i.d. entries drawn from the uniform distribution on $[0, 1]$ and h an arbitrary elementwise function with period 1.

Theorem 4.2 *Consider a set $\mathcal{X}$ of Q points in $\mathbb{R}^n$, where signals* $\mathbf{u}$, $\mathbf{v} \in \mathcal{X}$ *are measured using* (4.2) *to get* $\mathbf{y}$, $\mathbf{z}$. *Then, the following holds with probability greater than* $1 - e^{2 \log Q - 2m\varepsilon^2 \bar{h}^{-4}}$:

$$g(d) - \varepsilon \leq \frac{1}{m} \|\mathbf{y} - \mathbf{z}\|_2^2 \leq g(d) + \varepsilon \tag{4.3}$$

for all pairs $\mathbf{u}, \mathbf{v} \in \mathcal{X}$, *with a distortion g on the distances $d = d_{\mathcal{X}}(\mathbf{u}, \mathbf{v})$ equal to*

$$g(d) = 2 \sum_k |H_k|^2 (1 - \varphi_l(2\pi k|d)) \tag{4.4}$$

being $\bar{h} = \sup_t h(t) - \inf_t h(t)$, *$H_k$ the coefficients of the Fourier series of h and $\varphi_l(\xi|d)$ the characteristic function of the random variable denoting distance projected by a single row ϕ of matrix* $\mathbf{\Phi}$, *i.e.*, $l = \langle \phi, \mathbf{u} - \mathbf{v} \rangle$.

The most interesting thing that can be observed from Eqs. (4.3) and (4.4) is that the distortion on the distances can be controlled by a careful design of the function h and of distribution of the entries of the sensing matrix $\mathbf{\Phi}$. An important example is when the entries are i.i.d. Gaussian ($[\mathbf{\Phi}]_{ij} \sim \mathcal{N}(0, \sigma^2)$) and Euclidean distances are considered as $d_{\mathcal{X}}$. In that case, the distortion becomes

$$g(d) = 2 \sum_k |H_k|^2 (1 - e^{-2(\pi \sigma d k)^2}).$$

While the function h could in principle be anything, an important case is when it is a quantizer. Figure 4.7 shows an example with a periodic binary quantizer, which can be seen as an extension of sign random projections because when the period tends to infinity sign random projections are recovered. Suppose that i.i.d. Gaussians with zero mean and variance σ^2 are used for the sensing matrix and that the embedding is computed as

$$\mathbf{y} = Q(\mathbf{\Delta}^{-1}(\mathbf{\Phi u} + \mathbf{w})) \tag{4.5}$$

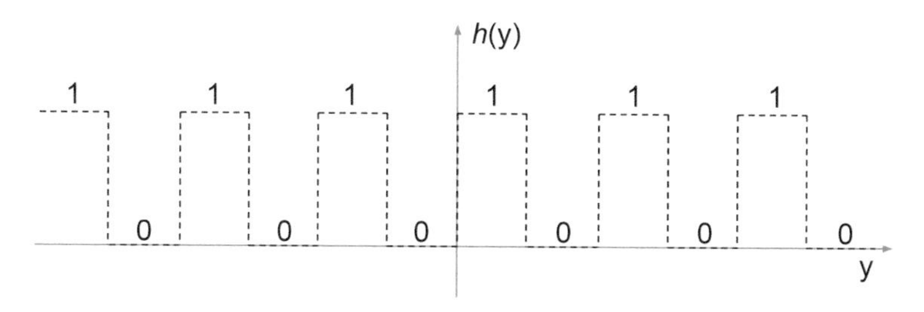

Fig. 4.7 Universal binary scalar quantizer

with $\boldsymbol{\Delta}$ being a diagonal matrix filled with scaling factor Δ, Q the function depicted in Fig. 4.7 with period 1, and $\mathbf{w}$ a dither uniformly distributed in [0, 1]. By specializing Eq. (4.4), the Hamming distance between the embeddings as a function of the Euclidean distance between the signals in the original space, $d = \|\mathbf{u} - \mathbf{v}\|_2$, can be computed to be

$$g(d) = d_H = \frac{1}{2} - \sum_{i=0}^{+\infty} \frac{e^{-\frac{\pi(2i+1)\sigma d}{\sqrt{2}\Delta}}}{(\pi(i+1/2))^2}.$$

Figure 4.8 shows the expected value of the Hamming distance in the embedded space with respect to the Euclidean distance in the original space for various values of the scaling factor Δ for a fixed σ. Notice that Δ essentially controls the period of the quantizer. It can be seen that for a large value of Δ the curve approaches the behavior of sign random projections, while for lower values it has a roughly linear region before saturating. This saturation phenomenon implies that only a range of distances is accurately embedded, while the geometry is distorted in such a way that the relationships between points lying at a distance that falls in the saturation region are lost. In [4], an information-theoretic argument is provided to show that no information about the relationships between such pairs of signals is available after the embedding. In particular, the mutual information between the embeddings is shown to be exponentially decreasing as a function of the distance in the saturation region.

Theorem 4.3 *Let* $\mathbf{u}, \mathbf{v} \in \mathbb{R}^n$ *and their universal binary embeddings* $\mathbf{y}, \mathbf{z} \in \{0, 1\}^m$ *computed as in* (4.5). *Then,*

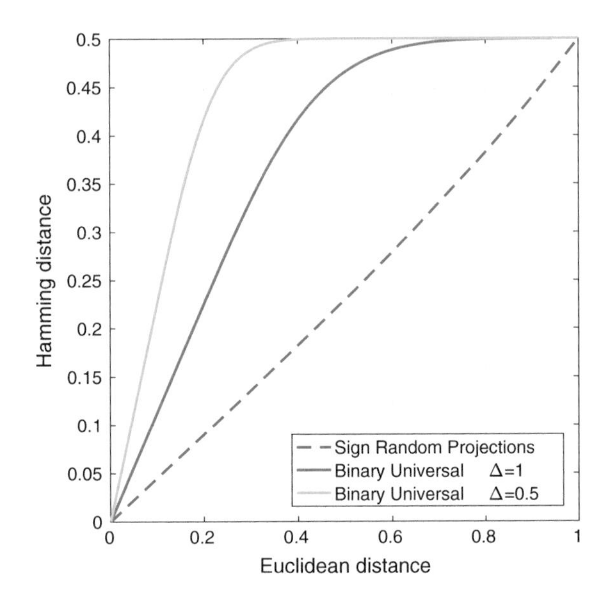

Fig. 4.8 Expected value of the Hamming distance between binary universal embeddings as a function of the Euclidean distance in the original space. Notice how the saturation region is larger when the period of the binary quantizer is smaller

$$I(\mathbf{y}; \mathbf{z}|d) \leq 10me^{-\left(\frac{\pi\sigma d}{\sqrt{2}\Delta}\right)^2}$$

being I the mutual information.

Comparison with Adaptive Embedding

In Sect. 4.1.2 we introduced the adaptive embedding as proposed in [29]. This embedding uses a reference signal to select a few random projections whose projection directions are more aligned with the reference. This introduces a bias in the conditional probabilities of bit-flip on the embeddings of the reference and of a test signal. This can be advantageous as it reduces the Hamming distances with respect to those achieved by sign random projections. Also, if we consider a binary classification problem where one class has large distances (e.g., it has zero mean correlation) and the other class has lower distances (and the reference is chosen from this class), the adaptive embedding can widen the distance gap between the two classes and therefore provide improved classification performance for the same dimensionality reduction.

Instead, notice that the universal embedding is significantly different because it is a non-adaptive approach. The geometric distortion operates in a different way. Instead of reducing distance values with respect to sign random projections, it expands them, up to creating a saturation region around $\frac{1}{2}$ bit flip probability. This is explained in the analysis in [6] where the theoretical results only hold for distance distortions that are subadditive functions, i.e., $g(a + b) \leq g(a) + g(b)$.

Figure 4.9 compares the Hamming distance in the embedded space with the angular distance in the original space (see Eq. 2.2). We can see that sign random projections have a linear behavior therefore their map satisfies subadditivity with the equality. The universal embedding is concave and therefore subadditive. On the other hand, the adaptive embedding is not. Notice that this is only possible because of the adaptivity and the distance is always evaluated between the embedding of the reference and the embedding of a signal.

4.2.3 *Private Clustering*

The goal of this section is to illustrate a problem where the privacy-preserving property of the universal embedding can be useful. Suppose an organization is surveying a number of people by posing a number of questions on sensitive matters. The answers are represented as a signal in $\mathbb{R}^n$ to be collected and sent to a centralized server for processing. As an assumption on the data, we know that they cluster in communities, where the distances between signals inside a community is small while the distances between communities are large, as depicted in Fig. 4.10 for $n = 2$. The goal for the server is to cluster the people by assigning them to the corresponding community. Due to the sensitive nature of the data, the server should not be able to have the original

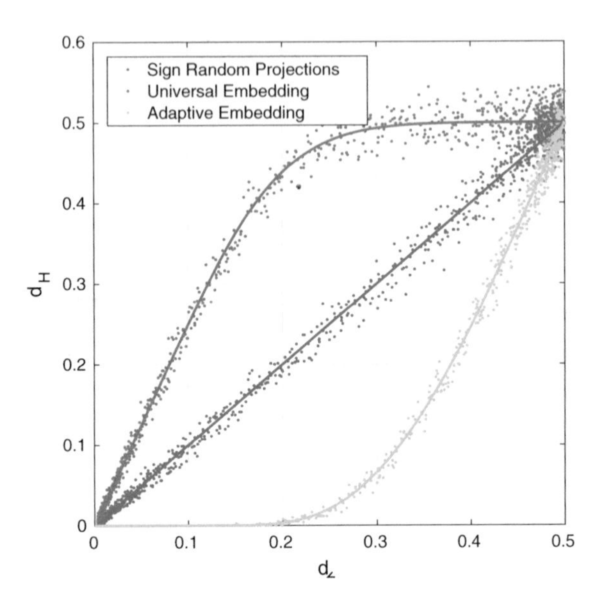

Fig. 4.9 Distance map from angular distance to Hamming distance for sign random projections, binary universal embedding, adaptive embedding

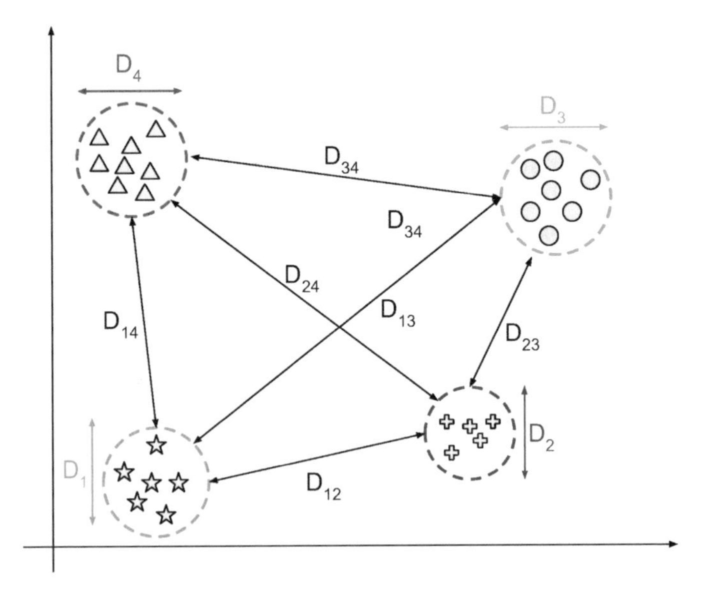

Fig. 4.10 Clustering problem. Multiple communities are present and such that the community diameter D_i is much smaller than the inter-community distances D_{ij}. The binary universal embedding can solve this problem with a bound on the maximum distance D such that $\max_i D_i \leq D \leq \min_{i \neq j} D_{ij}$

signal representation nor to infer the global geometry, i.e., estimate how much two communities are similar. This problem can be solved by universal embeddings.

A trusted third party is in charge of distributing the same sensing matrix to the people but not to the server. Also, a distance threshold is defined as the maximum diameter of a community, i.e., the maximum distance between two signals in a single community. This allows each person to compute a universal embedding of their signal and transmit it to the server. The server will then proceed to run a distance-based clustering algorithm on the embeddings, e.g., computing the nearest neighbors within a radius of each embedded signal.

The contents of the signals are hidden from the server thanks to the fact that the sensing matrix of the embedding is kept secret, using the same argument presented in Lemma 4.1.

On the other hand the full geometry is also hidden thanks to Theorem 4.3, according to which the server cannot infer anything about the relationships between pairs of signals whose original distances fell in the saturation region of the embedding.

References

1. Bianchi, T., Bioglio, V., Magli, E.: Analysis of one-time random projections for privacy preserving compressed sensing. IEEE Trans. Inf. Forensics Secur. **11**(2), 313–327 (2016)
2. Bianchi, T., Magli, E.: Analysis of the security of compressed sensing with circulant matrices. In: 2014 IEEE International Workshop on Information Forensics and Security (WIFS), pp. 173–178 (2014)
3. Bonneau, J., Herley, C., van Oorschot, P.C., Stajano, F.: Passwords and the evolution of imperfect authentication. Commun. ACM **58**(7), 78–87 (2015)
4. Boufounos, P., Rane, S.: Secure binary embeddings for privacy preserving nearest neighbors. In: 2011 IEEE International Workshop on Information Forensics and Security, pp. 1–6 (2011)
5. Boufounos, P.T.: Universal rate-efficient scalar quantization. IEEE Trans. Inf. Theory **58**(3), 1861–1872 (2012)
6. Boufounos, P.T., Rane, S., Mansour, H.: Representation and coding of signal geometry. Inf. Inference J. IMA **6**(4), 349–388 (2017)
7. Burr, W.E., Dodson, D.F., Newton, E.M., Perlner, R.A., Polk, W.T., Gupta, S., Nabbus, E.A.: Electronic authentication guideline. NIST Special Publication 800-63-2 (2013)
8. Caldelli, R., Amerini, I., Picchioni, F., Innocenti, M.: Fast image clustering of unknown source images. In: 2010 IEEE International Workshop on Information Forensics and Security, pp. 1–5 (2010)
9. Chen, M., Fridrich, J., Goljan, M., Lukas, J.: Determining image origin and integrity using sensor noise. IEEE Trans. Inf. Forensics Secur. **3**(1), 74–90 (2008)
10. Chierchia, G., Cozzolino, D., Poggi, G., Sansone, C., Verdoliva, L.: Guided filtering for PRNU-based localization of small-size image forgeries. In: 2014 IEEE International Conference on Acoustics, Speech and Signal Processing (ICASSP), pp. 6231–6235 (2014)
11. Dodis, Y., Ostrovsky, R., Reyzin, L., Smith, A.: Fuzzy extractors: how to generate strong keys from biometrics and other noisy data. SIAM J. Comput. **38**(1), 97–139 (2008)
12. Fridrich, J.: Digital image forensics. IEEE Signal Process. Mag. **26**(2), 26–37 (2009)
13. Goljan, M., Fridrich, J., Filler, T.: Large scale test of sensor fingerprint camera identification. In: Proceedings SPIE, Media Forensics and Security, vol. 7254, pp. 72,540I–72,540I–12 (2009)
14. Herder, C., Yu, M.D., Koushanfar, F., Devadas, S.: Physical unclonable functions and applications: a tutorial. Proc. IEEE **102**(8), 1126–1141 (2014)

15. Hinrichs, A., Vybíral, J.: Johnson-Lindenstrauss lemma for circulant matrices. Random Struct. Algorithms **39**(3), 391–398 (2011)
16. Juels, A., Wattenberg, M.: A fuzzy commitment scheme. In: Proceedings of the 6th ACM Conference on Computer and Communications Security, CCS'99, pp. 28–36. ACM, New York (1999)
17. Li, C.T.: Unsupervised classification of digital images using enhanced sensor pattern noise. In: Proceedings of 2010 IEEE International Symposium on Circuits and Systems, pp. 3429–3432 (2010)
18. Linnartz, J.P., Tuyls, P.: New shielding functions to enhance privacy and prevent misuse of biometric templates. In: International Conference on Audio-and Video-Based Biometric Person Authentication, pp. 393–402. Springer, Berlin (2003)
19. Lukáš, J., Fridrich, J., Goljan, M.: Detecting digital image forgeries using sensor pattern noise. Electronic Imaging, vol. 6072, pp. 60,720Y–60–720Y–11. International Society for Optics and Photonics, Bellingham (2006)
20. Lukas, J., Fridrich, J., Goljan, M.: Digital camera identification from sensor pattern noise. IEEE Trans. Inf. Forensics Secur. **1**(2), 205–214 (2006)
21. Mahdavifar, H., Vardy, A.: Achieving the secrecy capacity of wiretap channels using polar codes. IEEE Trans. Inf. Theory **57**(10), 6428–6443 (2011)
22. Rauhut, H.: Circulant and Toeplitz matrices in compressed sensing. In: SPARS'09 - Signal Processing with Adaptive Sparse Structured Representations (2009)
23. Valsesia, D., Boufounos, P.T.: Multispectral image compression using universal vector quantization. In: 2016 IEEE Information Theory Workshop (ITW), pp. 151–155 (2016)
24. Valsesia, D., Boufounos, P.T.: Universal encoding of multispectral images. In: 2016 IEEE International Conference on Acoustics, Speech and Signal Processing (ICASSP), pp. 4453–4457 (2016)
25. Valsesia, D., Coluccia, G., Bianchi, T., Magli, E.: Compressed fingerprint matching and camera identification via random projections. IEEE Trans. Inf. Forensics Secur. **10**(7), 1472–1485 (2015)
26. Valsesia, D., Coluccia, G., Bianchi, T., Magli, E.: Large-scale image retrieval based on compressed camera identification. IEEE Trans. Multimed. **17**(9), 1439–1449 (2015)
27. Valsesia, D., Coluccia, G., Bianchi, T., Magli, E.: User authentication via PRNU-based physical unclonable functions. IEEE Trans. Inf. Forensics Secur. **12**(8), 1941–1956 (2017)
28. Valsesia, D., Magli, E.: Compressive signal processing with circulant sensing matrices. In: 2014 IEEE International Conference on Acoustics, Speech and Signal Processing (ICASSP), pp. 1015–1019 (2014)
29. Valsesia, D., Magli, E.: Binary adaptive embeddings from order statistics of random projections. IEEE Signal Process. Lett. **24**(1), 111–115 (2017)
30. Vybral, J.: A variant of the Johnson-Lindenstrauss lemma for circulant matrices. J. Funct. Anal. **26**(4), 1096–1105 (2011)
31. Yin, W., Morgan, S., Yang, J., Zhang, Y.: Practical compressive sensing with Toeplitz and circulant matrices. Visual Communications and Image Processing, vol. 7744, p. 77440K. International Society for Optics and Photonics, Bellingham (2010)

Chapter 5
Conclusions

Compressed sensing has drawn remarkable attention because of its appealing compress-while-sampling property. Nonetheless, its noteworthy properties as privacy-preserving framework are among its lesser known aspects. This book enlightened this latter perspective by summarizing state-of-the-art results of CS as a cryptosystem and CS as a privacy-preserving embedding.

In Chap. 3 we showed that if the sensing matrix is considered to be a secret key, then CS can be equipped with some secrecy notions and act as a weak though efficient private-key cryptosystem. In more detail, a CS cryptosystem will always leak, at least, the energy of the plaintext unless additional methods are taken into account. Moreover, we showed that the use of structured or non-Gaussian sensing matrices decreases the secrecy of the whole system by increasing the amount of information which is leaked through the measurements. This is the case of practical systems making use of circulant, Bernoulli or quantized sensing matrices for which only a weaker notion of security can be provided. As a matter of fact, in order to consider CS cryptosystems a viable and effective alternative to standard private-key cryptosystems, the existing gap between practical systems and theoretical secrecy results has to be closed. In this regard, the characterization of the secrecy of cryptosystems making use of finite precision sensing matrices is still an open problem. This problem, has so far only been experimentally addressed in the case of quantized Gaussian sensing matrices, and further investigation is needed.

In Chap. 4 we presented signal embeddings and discussed how they can be used to solve inference problems with some privacy constraints. In particular, embeddings can be used to hide the signal content from a third party while still allowing this party to process them, e.g., for template detection or nearest neighbor search tasks. We discussed that this is possible thanks to the secrecy of the sensing matrix and the statistical properties of the embedded signals. Moreover, we also analyzed how certain embedding constructions allow to hide relationships between pairs of signals by conveniently distorting the geometry of the space.

M. Testa et al., *Compressed Sensing for Privacy-Preserving Data Processing*, SpringerBriefs in Signal Processing, https://doi.org/10.1007/978-981-13-2279-2_5

Printed in the United States
By Bookmasters